Prentice Hall Advanced Reference Series

Physical and Life Sciences

Genetic Transformation in Plants

R. Walden

PRENTICE HALL
ENGLEWOOD CLIFFS, NEW JERSEY 07632

Library of Congress Cataloging-in-Publication Data

Walden, R.
    Genetic transformation in plants / R. Walden.
        p.   cm.
    Includes bibliographies and index.
    ISBN 0-13-351040-9
    1. Plant genetic engineering. 2. Genetic transformation.
    3. Recombinant DNA.   I. Title.
    QK981.5.W34   1989
    581.1'5--dc19

                                                        89-3575
                                                        CIP

North American English-language edition
published 1989 by Prentice-Hall, Inc.
A Division of Simon & Schuster
Englewood Cliffs, New Jersey 07632

Prentice Hall Advanced Reference Series

The publisher offers discounts on this book when ordered
in bulk quantities. For more information, write:

> Special Sales/College Marketing
> Prentice-Hall, Inc.
> College Technical and Reference Division
> Englewood Cliffs, NJ 07632

Printed in the United States of America

10 9 8 7 6 5 4 3 2 1

ISBN 0-13-351040-9

Dedicated to my family

Contents

Preface and Acknowledgements

The recent advances made in the techniques of transfer of foreign DNA into plant cells have been dramatic. It has now become possible to transfer any gene to the plant cell, have it stably integrated into the plant genome and arrange for it to be expressed, if required, in a developmentally controlled manner. Plant transformation is likely to become an indispensable aid to both the plant physiologist and biochemist in understanding the role that individual gene products might play in the life of the plant. In addition these developments have aroused much interest in the possibility of exploiting recombinant DNA technology in crop improvement. Due to the rapidity of the advances made in this area of research it has been difficult for the non-specialist to keep abreast of the techniques and results. Generally, the information available has been confined to the original scientific report or has appeared in conference proceedings. Hence, in writing this book I have attempted to muster all the information that has been obtained concerning plant transformation and the expression of foreign genes in plants in an attempt to provide, for both the specialist and the non-specialist alike, a review of this exciting area of research. The aim of the book is to introduce the background to plant transformation and describe what has been achieved, where the limitations of our current technology may lie and what might be achieved in the near future. While it is intended that this book will provide a detailed summary of plant transformation for the molecular biologist I hope that it will also serve as an introduction to physiologists and biochemists so that they can appreciate the potential that gene transfer might have for their own research interests. Throughout I have assumed that the reader will have a basic knowledge of both recombinant DNA technology and plant biology. I have attempted to avoid 'jargon' as much as possible and have included a glossary at the end of the book in an attempt to clarify subjects raised in the text.

The techniques of plant transformation involve both molecular biology and plant tissue culture. In writing this book I have concentrated on the former area because it is where my own personal interest lies and the fact that there is another title in this series of books which is dedicated to plant tissue culture. Where possible I have concentrated on the production of transgenic plants and the expression of foreign genes in them, hence I do not discuss in detail the important area of using transgenic plant cells in culture to produce not only secondary products but also polypeptides.

In summarizing broad subject areas I have limited references to reviews where the reader can obtain further information. When discussing recent results I have attempted to include as many references to original research articles as possible so that the reader can refer directly to the original data. In addition I have confined the work that is described to that which has appeared in the scientific literature. Size constraints have seriously limited the number of references that I have been able to include and I have, by no means, been able to compile an exhaustive review of the literature. The results that are described in this book have been derived from many different laboratories and frequently several workers have been carrying out similar pieces of research simultaneously. I have tried as much as possible to give research credit where appropriate. To those that I have not been able to credit, or have missed, my sincere apologies. The text of this book was completed in January 1988 and I hope that it is a comprehensive review of the work that has been published to that date.

I have avoided writing a 'how to' book, largely because there are several useful books appearing in this area. For those who wish to find more practical details of how to transform plants I would recommend *Plant Genetic Transformation and Gene Expression: A Laboratory Manual* edited by John Draper and others which is adapted from the manual produced for the Plant Transformation and Gene Expression Course run by the Department of Botany and the Biocentre at Leicester University. For those wishing to find out more about plant molecular biology in general I would recommend *Plant Molecular Biology* by Don Grierson and Simon Covey published by Blackie.

In writing this book I have avoided three areas: the ethics of gene transfer, whether gene transfer will be of significant benefit to agriculture particularly in the Third World, and the biosafety concerns that are raised by some of the work that is described. These topics merit a book in themselves and we ignore them at our peril.

I'd like to thank everyone at the Leicester Biocentre who has given me support during the year that I have taken to write this book, particularly the members of the Plant Molecular Biology group: Roy Dunford, Helen Thompson, Ian Loft, Dianne Firby and Sabine Rosahl. Dianne and Sabine, with John Bryant, deserve special thanks for reading drafts of the text and being constructive in their criticisms. My thanks also to Dianne, Rod Scott, Jurek Paskowski, Liz Dennis, Nilgun Tumer and Dilip Shah who provided me with photographs for the text.

Figure Acknowledgements

The following figures have been reproduced by kind permission of the copyright holders.

Chapter 5

Fig. 5.1   Paskowski, J., Shillito, R.D., Saul, M., Mandak, V., Hohn, T., Hohn, B. and Potrykus, I. (1985). *EMBO J.* **3**, pp. 2717–2722. Oxford, IRL Press Limited.
Fig. 5.2   Howard, E.I., Walker, J.C., Dennis, E.S. and Peacock, W.J. (1987). *Planta* **170**, pp. 535–540. Heidelberg, Springer-Verlag.

Chapter 7

Fig. 7.1   Shah, D., Horsch, R.B., Klee, H.J., Kishore, G.M., Winter, J.A., Tumer, N., Hironaka, C.M., Sander, P.R., Gasser, C.S., Aykent, S., Siegel, N.R., Rogers, S.G. and Fraley, R.T. (1986) *Science 233*, pp. 478–481. Copyright 1986 by the AAAS.
Fig. 7.4   Tumer, N., O'Connell, K.M., Nelson, R.S., Sanders, P.R., Beachy, R.N., Fraley, R.T. and Shah, D.M. (1987). *EMBO J.* 6, pp. 1181–1188. Oxford, IRL Press Limited.

Genetic Transformation in Plants

Chapter 1

Conventional and Non-conventional Gene Transfer in Plants

Conventional Plant Breeding

Broadly speaking, the objectives of plant breeding are to improve the yield, the quality or the reliability of a particular crop plant. Breeders have to bear in mind that the farmer wants a crop that he can grow consistently to make maximum profit and in order to do this the crop must reach the market requirements of either governments, commodity dealers or the consumer. Faced with this the plant breeder is interested in introducing genetic diversity into plant populations by intercrossing or mating appropriate germplasm which contains useful characteristics able to complement those already present in the crop. Plants which display an improved phenotype and hence contain the genes responsible for the desired trait are selected. Once this has been achieved, subsequent growth and selection is carried out until genetic uniformity, agronomic stability and maximum reproductive ability are obtained. Often a variety of locations are used so that plant performance can be assessed in different geographical and climatic conditions. In order to undertake this process the breeder has to take into account two important aspects of plant biology. Firstly, the method by which plants can be multiplied, either by vegetative propagation or by seed, whether by self- or open pollination. This is obviously important in determining the strategy to be used in breeding and how the resultant material can be multiplied. Secondly, the parts of the plant that form the end product, whether these are seeds, fruits, leaves, stems, roots, tubers or buds. The final economic performance of the product is the major consideration in breeding although the characteristics of the individual crop which are sought by the breeder may vary considerably and include such diverse considerations as the timing of seedling emergence from the seed, seedling vigour, the stature of the plant at harvest, ease of harvest and the storage properties of the harvested product.

Hence, the plant breeder uses his skill to manipulate a wide variety of characteristics and although some of these may be common to a variety of crop plants, this is by no means always the case. Moreover, although breeding depends on plant genetics, the ideals that are often sought are likely to be multifaceted, resulting from the expression of a number of genes in response to a wide variety of epigenetic factors.

At the molecular level, plant breeding depends on mutation and recombination, whereas at the practical level it depends on the judgement of the breeder to select the plant with the appropriate phenotype. It is hard to say when this form of 'genetic engineering' started. Indeed it is arguable that it was initiated as soon as the plant became a genetic entity, as natural selection. Although human intervention has accelerated the process dramatically, this involvement has been relatively recent, in evolutionary terms starting possibly between 9000 and 11000 years ago. The work of Darwin and the rediscovery of Mendel's laws in 1900 provided the impetus to place plant breeding on a firm scientific footing and since then there have been dramatic advances in crop improvement. In plant breeding the breeder may use either interspecific gene transfer, the transfer of genes from a non-cultivated plant species to a crop variety in a related species, or intergeneric transfer where transfer is attempted between a wild species to domesticated relatives either in the same or even different genera. However although a great deal of success has been achieved by these techniques, difficulties are often encountered in transferring genes between species because of natural barriers which result in reductions in fertility after genetic transfer.

Tissue Culture as a Tool to Produce Novel Hybrids

The development of plant tissue culture techniques in the 1960s and 1970s began to provide the means by which some of the natural barriers to gene transfer could be overcome. For example, reduction in fertility following crosses between species which results from the cessation of embryo development because of inadequate nutrient supply can be overcome by 'embryo rescue' where the immature embryo is isolated and allowed to develop *in vitro* in the presence of the appropriate nutrients and growth hormones. Moreover, the observation that isolated plant cells were totipotent and could regenerate to produce fully differentiated plants formed the practical foundation for the concept of gene transfer in plants by non-sexual means. One way in which this might be carried out is by the fusion of isolated plant cells to produce somatic hybrids. Protoplasts isolated from different species can be induced to fuse on either incubation with chemicals such as polyethylene glycol or in the presence of an electric current. The resultant somatic hybrid can then be grown *in vitro* to produce an undifferentiated mass of cells, known as callus, from which, in some species, plants can be regenerated. Much interest has been paid to the production of novel hybrids by this method but, despite this attention, success has been limited largely as a result of three things: (i) the inability to regenerate plants from a large number of species of agronomic importance, (ii) the infertility of some somatic hybrids and (iii) the relatively precise recombination required to exchange genetic material between the parental nuclear genomes so that chromosomal aberrations, which may affect the stability of the genome, are avoided. Some success has, however, been achieved in the related technique of producing 'cybrids'. These are the products of cell fusion experiments where the cells contain the nuclear DNA of one parent and the cytoplasmic genomes of the chloroplast and the mitochondria from the other parent. 'Cybridisation' has been used successfully to engineer cytoplasmic male sterility and herbicide resistance (see review by Evans 1983). Nevertheless this technique is still limited to the number of useful traits that might be encoded by the organelles. The difficulties encountered by engineering the traits of plants by the production of somatic hybrids or cybrids might be overcome to a certain extent by the application of recombinant DNA technology. In addition this also raises the possibility of transferring a single, well-defined gene encoding a specific trait to the plant genome. In so doing, the difficulty often encountered in conventional breeding of a desired characteristic being co-inherited with a detrimental character from which it cannot be simply removed by subsequent backcrossing, may be overcome. As our knowledge of plant molecular biology has increased this approach has become an attractive possibility.

The Advent of Plant Transformation

The 1960s and 1970s witnessed an explosion of our knowledge of gene structure and function. We became aware of not only what a gene looked like and what controlled its expression but also how one or more genes could be transferred from one cell to another. Initially this knowledge was based primarily on bacterial systems, but increasingly attention has been turned to a wide variety of cell types including those of animals and plants. Concomitantly, recombinant DNA technology has allowed us to isolate, dissect and re-introduce specific gene sequences into a great many different host cells and in many cases the foreign DNA can function normally in the novel environment. These developments, coupled with the advances in plant tissue culture, encouraged the idea that useful agronomic traits might be introduced into crop plants by the transfer of an isolated, well-characterized sequence of DNA. This notion led to a flurry of reports on experiments carried out with the aim of demonstrating the uptake of DNA into plant tissues, isolated protoplasts or pollen.

Routinely, in the first attempts to transform plants, the DNA used was isolated from bacteria and the aim was to demonstrate that it could enter the plant cell and integrate into the genome or replicate autonomously in the cell. In the former type of experiment, radioactively-labelled donor DNA, which has a different density from plant DNA, was applied to plant tissues or protoplasts. Following an incubation period total nucleic acid was extracted and subjected to CsCl density gradient centrifugation with the resultant banding of the DNA being used to infer whether the DNA had entered the plant cell and had been integrated into the plant genome. The experiments carried out to demonstrate replication of the foreign DNA were similar, but in this case the DNA that was applied was not radioactively labelled. Following an incubation period radiolabelled thymidine was applied with the expectation that it would be integrated into DNA that was undergoing replication which could be subsequently resolved by CsCl density gradient centrifugation. The difficulty with these types of experiment was that the results obtained were also consistent with possible bacterial contamination and the presence of residual bacterial DNA binding to the outside of either the plant cell or the nuclear membrane. In some cases the DNA might have been taken up into the cell but it was not clear whether it had been maintained as a functional unit or had been metabolized to produce precursors for DNA synthesis. Hence in these cases evidence to suggest that plant transformation had taken place has been summarized as being 'at best tentative' (Kleinhofs and Behki 1977).

One way of circumventing these difficulties is to attempt to demonstrate the biological activity of the foreign DNA in the plant cell. This is most easily accomplished by using a genetic marker. The initial attempts to do this utilized bacteriophage containing a specific marker gene to complement a mutant cell line. Examples of this imaginative approach involved the inoculation of plant cell lines with bacteriophage containing a bacterial β-galactosidase gene (Doy *et al.*, 1973; Johnson *et al.*, 1973). In these cases the control cells were unable to grow on lactose but, following inoculation with the bacteriophage, they were able to grow, undergo limited amounts of division and exhibited a transient increase in β-galactosidase activity following transfer to lactose. The drawback with this type of experiment was in being able to demonstrate that the foreign DNA was within the cell and not attached to the outer cell membrane. Moreover, there was no definitive evidence that the β-galactosidase was expressed within the plant cell and it was also possible that the cells themselves might, at very low frequency, adapt to growth on lactose (Kleinhofs and Behki 1977).

Although these experiments aroused a great deal of interest, they did not prove to be widely accepted and this was due largely to the limitations of the experimental systems that were currently available and in particular the inability to demonstrate conclusively that transformation by foreign DNA had taken place. Indeed, at this time experimenters were faced with the double dilemma of not only attempting to show that the DNA had entered the plant cell but also hoping that the foreign DNA once within the cell would function normally. It is perhaps significant that the majority of this work was carried out before a single plant gene encoding a polypeptide had been isolated, cloned or identified, a feat eventually achieved in 1977 with the localization of the gene encoding the large subunit of ribulose bisphosphate carboxylase on the chloroplast genome of maize (Coen *et al.*, 1977). Hence, to a certain extent, the aspirations of those involved in the initial attempts of plant transformation preceded our detailed knowledge of the molecular basis of the control of gene expression and the involvement of the gene products in the biochemistry of the plant cell.

Development of Strategies for Plant Transformation

More recently, the extremely sensitive techniques of Southern and Northern blot analysis, which can be used to demonstrate that a particular DNA or RNA is present in the plant cell, have become available. In addition, attention has become focussed on the development of vectors which can be used to transfer specific regions of DNA into the plant cell. These can be based potentially on either elements derived from the plant genome or from the DNA (or RNA) of plant pathogens (for a review see Howell 1982). As is discussed in subsequent chapters, most success in plant transformation has been achieved using DNA derived from plant pathogens. Indeed the mechanism utilized by one pathogen, *Agrobacterium*, to transform the plant cell has been used most extensively to transfer a wide variety of foreign DNAs into the plant genome. Moreover, genetic markers have become available which confer a new phenotype on the transformed plant cell, thus allowing it to be isolated by direct selection. These routinely consist of genes from bacteria conferring resistance to an antibiotic, linked to DNA sequences which can direct their correct expression in plant cells. This then allows the recovery of transformants containing foreign DNA from a large number of non-transformed cells. The combination of vector development and the availability of genetic markers now allows the conclusive identification of a transformed cell or whole plant. Routinely, a number of criteria are used and can include a phenotypic change, the demonstration of the presence of foreign DNA in the plant cell, the presence of RNA corresponding to the sequence of foreign DNA, a unique enzyme activity and the stable inheritance of the foreign DNA.

We shall see that there are now a variety of ways of introducing DNA into the plant cell and obtaining stably transformed plants. Indeed the transformation of a few plant species has become a simple laboratory exercise raising many exciting possibilities. On the one hand, novel traits considered to be agronomically important have been introduced into crop plants, whilst on the other hand it is likely that plant transformation will play a central role in increasing our understanding of both plant gene expression and biochemistry (Schell 1987).

References

Coen, D.M., Bedbrook, J.R., Bogorad, L. and Rich, A. (1977). 'Maize chloroplast DNA fragment encoding the large subunit of ribulose bisphosphate carboxylase', *Proc. Nat. Acad. Sci. USA* **74**, pp. 5487-5491.

Doy, C.H., Gresshoff, P.M. and Rolfe, B.G. (1973). 'Time-course of phenotypic expression of *Escherichia coli* gene Z following transgenosis in haploid *Lycopersicon esculentum* cells', *Nature (New Biology)* **244**, pp. 90–91.

Evans, D.A. (1983). 'Agricultural applications of plant protoplast fusion', *Bio/Technology* **1**, pp. 253–261.

Howell, S.H. (1982). 'Plant molecular vehicles: potential vectors for introducing foreign DNA into plants', *Ann. Rev. Plant Physiol.* **33**, pp. 609–650.

Johnson, C.B., Grierson, D. and Smith, H. (1973). 'Expression of λplac5 DNA in cultured cells of a higher plant', *Nature (New Biology)* **244**, pp. 105–106.

Kleinhofs, A. and Behki, R. (1977). 'Prospects for plant genome modification', *Ann. Rev. Genet.* **11**, pp. 79–101.

Schell, J. (1987). 'Transgenic plants as tools to study the molecular organization of plant genes', *Science* **237**, pp. 1176–1183.

Chapter 2

The Biology of Tumour Formation by Agrobacterium

Introduction

Agrobacterium tumefaciens has been called a natural genetic engineer of plants. The mechanism by which it, and its close relative, *Agrobacterium rhizogenes*, induce tumour formation and hairy root disease, respectively, on host species is arguably one of the most fascinating areas of plant biology. Galls or tumours can be induced on plants by a variety of agents such as insects, viruses and fungi. They can be classified as either self-limiting or non-self-limiting depending on whether the presence of the inducing agent is continuously required for the changed phenotype. For example, clubroot disease of brassicas requires the continued presence of the plasmodia of *Plasmodiophora brassicacae* for its proliferation whereas genetic tumours arising in interspecific hybrids of plants occur spontaneously on mature tissue without any apparent external induction and are thought to arise from the induction of aberrant gene activity. Tumour induction by *Agrobacterium tumefaciens* has received a great deal of attention because initially it was thought that this might shed light on the mechanism of cancer formation, although more recently much interest has been paid to the system because of the possibility of using the Ti plasmid of *Agrobacterium tumefaciens* as a vector for the transformation of plant cells. This chapter provides a brief introduction to tumour formation by

Agrobacterium and then describes in detail our current knowledge of the mechanism by which it takes place. This area has been reviewed in detail fairly recently (Kahl and Schell 1982), hence an exhaustive review of the large amount of literature that has been published on this topic will not be given, but rather a background to the biological system highlighting some of the points that need to be considered when using the Ti plasmid, or the Ri plasmid of *Agrobacterium rhizogenes*, as a plant transformation vector.

Tumour Formation by *Agrobacterium tumefaciens*

Crown gall, which has been recognized as a neoplastic disease for almost 90 years, is characterized by disorganized growth on the stem of plants near the surface of the soil (Fig. 2.1). Normally galls form on a variety of dicotyledonous species although recently it has been demonstrated that tumours can also form on some monocotyledonous plants. At the beginning of this century it was found that tumours could be induced in plants following inoculation of a pure culture of bacteria that had been isolated from tumourous tissue. It was thought that crown gall was analogous to the malignant tumours seen in animals and should be studied as such. This view gained support from the observation that tumours could be serially transplanted from plant to plant and continue to grow in a manner similar to transplantable animal tumours. By grafting tumour tissue from

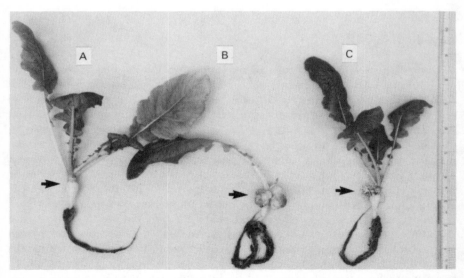

Fig. 2.1 Tumour formation by *A. tumefaciens* and *A. rhizogenes*. Three-week-old turnip plants were inoculated at the soil surface (position of arrow) following wounding. Photograph taken three weeks later. (A) Uninoculated; (B) inoculated with *A. tumefaciens*; (C) inoculated with *A. rhizogenes*.

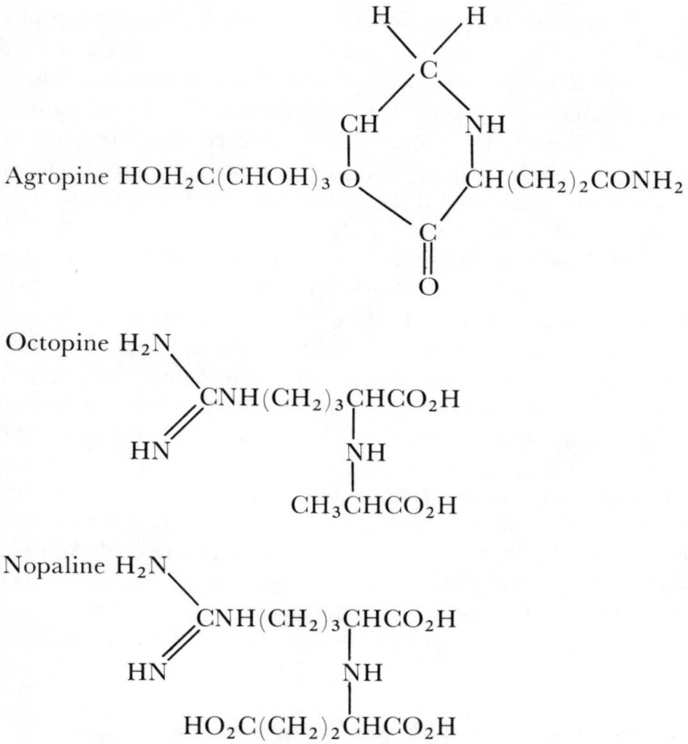

Agropine $HOH_2C(CHOH)_3$

Octopine

$$H_2N$$
$$CNH(CH_2)_3CHCO_2H$$
$$HN \qquad NH$$
$$CH_3CHCO_2H$$

Nopaline

$$H_2N$$
$$CNH(CH_2)_3CHCO_2H$$
$$HN \qquad NH$$
$$HO_2C(CH_2)_2CHCO_2H$$

Fig. 2.2 Structures of three opines.

sugar beet onto either fodder beet or red beets and *vice versa* it was found that the colour of the beet could be used as a marker to demonstrate that tumours arising from transplanted tumours arose entirely from the implanted tissue. Further work showed that as tumours aged the inciting bacterium, which was at first called *Bacterium tumefaciens* but was later to be named *Agrobacterium tumefaciens*, tended to die and it became impossible to isolate bacteria from grafted tumours. Hence, it appeared that after tumour induction the tumour could proliferate in the absence of the inciting bacterium. This indicated that the phenotype of the tumour cell was permanently transformed.

In the 1930s it was found that a dried extract from the culture medium in which the tumour-inducing bacterium had grown could induce in plant explants growth which resembled the tumours induced by the bacterium. The active substance was identified as indole-3-acetic acid (IAA) which came to be considered as an auxin. Subsequent work demonstrated that the bacterium also secretes the cytokinins *trans*-zeatin and/or isopentyladenine. The biological importance of these substances was demonstrated by the pioneering work on plant tissue culture carried out in the late 1950s which showed that the enlargement and division of isolated tobacco pith cells required the presence of an auxin and a cytokinin in the growth medium.

When auxin and cytokinin are added to a growth medium in a defined ratio, undifferentiated growth of plant tissue occurs, giving rise to callus. If the ratio of auxin to cytokinin is high, roots form; when this is reversed and the ratio of cytokinin to auxin is high, the growth of buds is initiated from which shoots develop. Whereas normal untransformed tissue requires auxin and cytokinin for continued proliferation *in vitro*, tumour tissue does not. This is considered a general characteristic of neoplastic growth and can be used to select for transformed tissue.

The observation that certain compounds were necessary for the continued growth and division of plant cells in tissue culture and that these were not necessary for the continued proliferation of tumour tissue, coupled with the observation that tumour tissue was able to grow in the absence of the inducing bacteria, suggested that tumour cells were able to continuously synthesize plant growth factors. As described later, this is indeed the case. Moreover, the tumour cell produces novel amino acid and sugar derivatives collectively called opines (Fig. 2.2). The type of opine (for example nopaline, octopine, agrocinopine and agropine) is dependent on the strain of *Agrobacterium* that induced tumour formation. The assay for opine synthesis can be carried out with relative ease (Chapter 3), hence the presence of opines has often been used as a chemical marker for transformed tissue.

Tumour Formation is Associated with the Presence of a Plasmid in *Agrobacterium*

There is a very narrow temperature range between 28 and 30°C for tumour induction. The reason for this remains unknown but it has been found that the transfer of plasmids during bacterial conjugation was also limited to the same temperature range and it has been suggested that there may be a thermosensitive step common to both processes. The causative agent in tumour formation was found to be lost from the bacterium when grown at 36°C. Subsequently it was demonstrated that the loss of virulence was accompanied by the loss of a large plasmid, the tumour-inducing or Ti plasmid. The Ti plasmid is a double-stranded circular DNA of approximately 200 kilobases (kb) in size. Not only are Ti plasmids responsible for tumour formation but they also specify the type of opine synthesized by the tumour cells. The opines induce conjugational transfer of the Ti plasmid between bacteria and the Ti plasmid itself encodes functions allowing the uptake and catabolism of the opine whose synthesis it has induced, enabling the bacterium to use it as a source of carbon and nitrogen. Because the bacterium which induces the tumour can use the opines as an energy source, it is thought that this confers on it a selective advantage over other soil bacteria so that, in effect, *Agrobacterium* can be considered to be a sophisticated parasite.

The Physical Basis of Tumour Formation

The work described above suggested that tumour cells were genetically transformed by *Agrobacterium* and this notion was proved correct by the analysis of the nuclear DNA of tumour cells which indicated that particular fragments of the Ti plasmid were associated with it. Detailed Southern blot analysis revealed that a specific segment of the Ti plasmid, the transferred or T-DNA, was integrated into the genome of the plant cell and that it was this that determines the tumourous phenotype. The T-DNA found in the plant cells is colinear with the T-DNA present in the Ti plasmid indicating that no major rearrangement of the sequence takes place during the transformation process. The sequences of plant DNA that flank the T-DNA have been isolated from a large number of transformants and studied by both restriction enzyme analysis and hybridization back to the plant genome and it is generally considered that the T-DNA inserts at random into the plant nuclear DNA.

One or more copies of the T-DNA can be present in the transformed plant genome but generally the borders of the T-DNA in the plant genome are delimited by a nearly perfect direct repeat sequence of 25 base pairs (bp) (TGGCAGGATATATTC(*or* G)XG(*or* A)T(*or* G)TGTAAA(*or* T)T(*or* C)) which also flank the T-DNA in both nopaline and octopine Ti plasmids. The right border repeat is required for the efficient transfer of DNA to the plant cell whereas the left border repeat is not. The right border repeat can be substituted by either a left border repeat or a synthetic border provided that they are in the correct orientation.

The Functional Organization of the Ti Plasmid

Ti plasmids found in different strains of *Agrobacterium* have four regions of homology as judged by Southern blot analysis and heteroduplex mapping. Two of these, the T-DNA and the *vir* region, have been shown by genetic analysis to be directly involved in tumour formation whereas the other two contain the regions that control plasmid replication and encode functions concerned with conjugative transfer (Fig. 2.3). The T-DNA and the *vir* regions from different Ti plasmids show regions of homology and the functions of both regions on one plasmid can often be complemented by those on another plasmid. In nopaline strains (e.g. C58, T37) the size of the T-DNA is approximately 24 kb. Octopine strains (e.g. B6S3, Ach5) contain two regions of T-DNA, T_L and T_R of 14 and 7 kb, respectively. T_L is present in all transformed cell lines and is functionally equivalent to the right portion of the nopaline T-DNA whereas T_R is not always present in tumour tissue; when it is, it may not be contiguous with T_L and its copy number can differ, suggesting an independent transfer of the two segments of DNA.

To the right of the right T-DNA border of octopine Ti plasmids is a 24 (bp) sequence called 'overdrive' (GAGCTCGTGAATAAGTCGCTGGTGTATGT-TTGTTTG) which is required for optimal T-DNA transfer. Overdrive sequences appear to act in *cis*, although their distance from, and orientation with respect to,

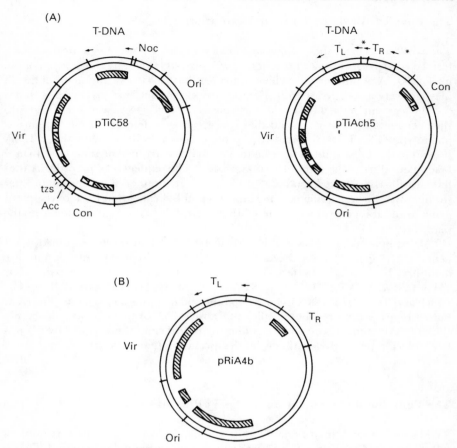

Fig. 2.3 Maps of the Ti and Ri plasmids. (A). Nopaline-type pTiC58 and octopine-type pTiAch5: hatched boxes indicate regions of sequence similarity between the two plasmids. (B). pRiA4b: hatched boxes indicate regions of sequence similarity with pTiT37 and pTiA6. Major functional regions are: Vir, virulence; Ori, origin of replication; Con, region important in the conjugative transfer of the plasmid; Noc, nopaline catabolism; Acc, agrocinopine catabolism; *tzs*, *trans*-zeatin synthesis and the T-DNA. Arrows indicate the positions of the T-DNA border repeats and overdrive sequences are marked with an asterisk, where known.

the right border can vary. As yet, the function of overdrive is not known but it appears to be important in the production of T-DNA transfer intermediates during the initial stages of plant cell transformation.

Transcript mapping of the T-DNA of nopaline Ti plasmids has revealed 13 transcripts whereas sequencing of the T_L and T_R DNA of octopine Ti plasmids has revealed 8 and 6 open reading frames, respectively. The sequences of the genes of

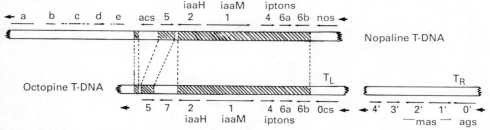

Fig. 2.4 Functional organization of the nopaline and octopine T-DNA. Short arrows indicate border repeat sequences, long arrows with letters and numbers indicate different transcripts.

the T-DNA and the regions that flank them are reminiscent of those found in eucaryotic genomes. This feature raises the interesting question of how they evolved to be present in a procaryotic organism. The nopaline T-DNA and the octopine T_L DNA share regions of extensive homology known as the 'core' region which contains the genes that encode the proteins involved in hormone biosynthesis, nopaline (or octopine) secretion and tumour size (Fig. 2.4).

A variety of techniques have shown that the T-DNA encodes enzymes responsible for the synthesis and secretion of opines by the plant cell as well as the enzymes involved in the biosynthesis of the hormones which play a major role in the establishment and maintenance of the tumour phenotype. The genes for nopaline synthase (*nos*) or octopine synthase (*ocs*) reside near the right border sequence whereas the gene for agrocinopine synthase (*acs*) is almost in the middle of the nopaline T-DNA. Mutation analysis indicates that nopaline or octopine secretion by the plant cell (*ons*) is encoded by gene 6a (Messens *et al.*, 1985). Mutations in genes 1 and 2 (*tms1* or *iaaM* and *tms2* or *iaaH*) lead to the formation of slow growing tumours which subsequently form shoots whose growth can be suppressed by the addition of auxins. This suggests that these genes encode proteins which produce compounds functionally equivalent to auxins. Conversely, mutations in gene 4 (*tmr* or *ipt*) give rise to slow growing tumours which frequently produce roots and root formation can be suppressed by the addition of cytokinins suggesting that this gene encodes a function giving rise to a cytokinin-like compound. Using the expression of the genes in *E. coli* mini-cells to over-produce proteins and assaying their activity with antibodies raised against IAA, as well as mutation analysis, it has been shown that gene 1 encodes tryptophan monooxygenase which converts tryptophan to indole-3-acetamide and gene 2 encodes indole acetamidehydrolase which converts indole-3-acetamide to IAA (Fig. 2.5A). The same pathway is used by another plant pathogen, *Pseudomonas savastonoi*, which produces galls on olive and oleander. Although the precise pathway for cytokinin synthesis in plants is not known, it is thought that the primary step involves a condensation reaction between 5'AMP and Δ^2-isopentenyl pyrophosphate (Fig. 2.5B) and it has been found that gene 4 encodes an isopentylenyl transferase activity which can carry this out. Gene 6b

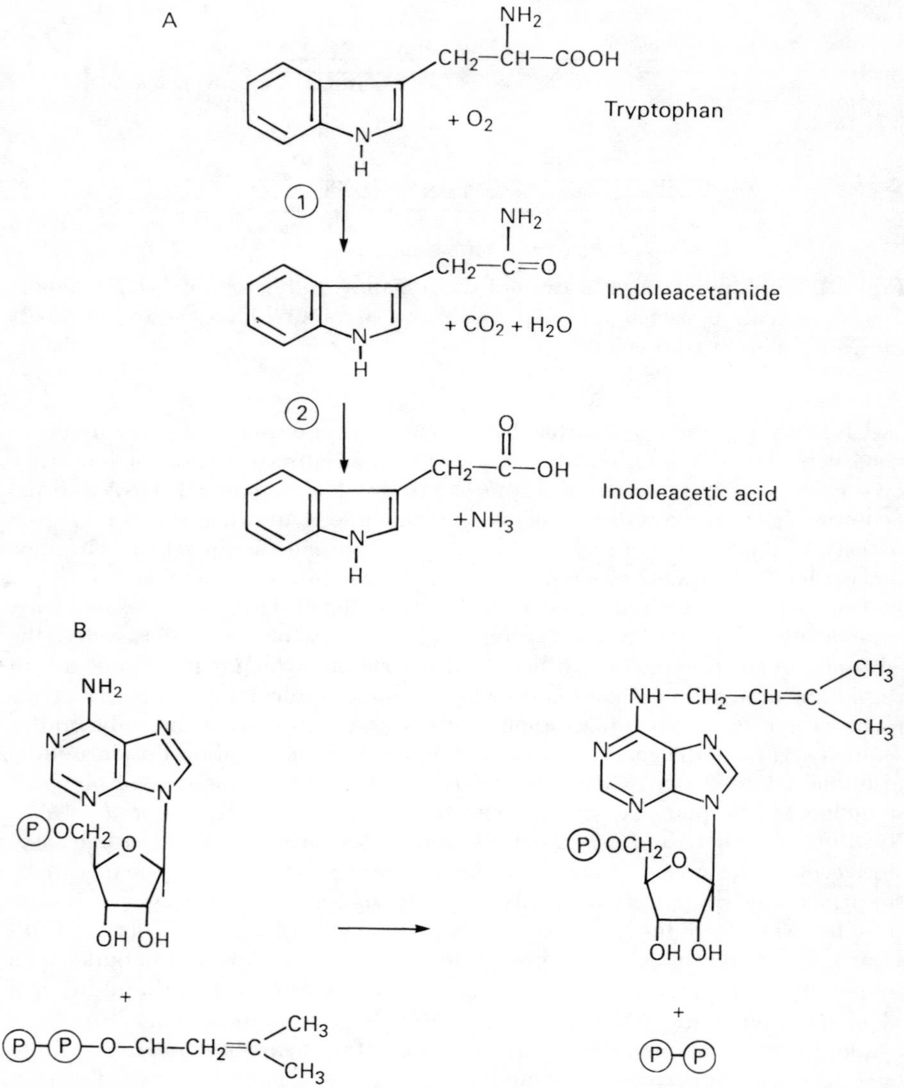

Fig. 2.5   Biosynthetic pathways for auxin and cytokinin encoded by the T-DNA. (A). Auxin biosynthesis involves the conversion of tryptophan to indoleacetamide and then to indoleacetic acid by tryptophan monoxygenase (1) and indoleacetamide hydrolase (2). (B). Cytokinin synthesis involves the attachment of an isopentenyl group to 5′ AMP by isopentyl transferase.

(*tml*) by a mechanism that is not known, is involved in the control of the size of the tumour on certain host species because mutations in this gene result in the formation of large tumours. The genes that are responsible for the tumour phenotype are, in the classic sense, oncogenes and hence are often referred to collectively as the *onc* genes. The T_R region of octopine T-DNA is not necessary for tumour formation and encodes enzymes for the synthesis of additional opines, transcript 1' and 2' being required for mannopine synthesis and transcript 0' is necessary for the conversion of mannopine into agropine (Salomon *et al.*, 1984).

Deletion analysis has shown that the genes encoded by the T-DNA can be replaced without interfering with the transfer of the DNA to the plant cell as long as the 25 bp border sequences are maintained. Moreover, T-DNA from which the oncogenic genes have been removed, once integrated into the plant genome, can be inherited in a Mendelian fashion suggesting that the foreign DNA is stable in the genome.

The *vir* genes of both nopaline- and octopine-type Ti plasmids are organized into at least six conserved complementation groups, *virA, B, C, D, E* and *G*. Octopine plasmids contain an additional complementation group, *virF*. Mutations in many of the *virA, B, G* and *D* genes abolish tumour formation whereas mutations in *virC, E* and *F* do not, indicating that they are not absolutely required and may be involved in the efficiency of transfer of the T-DNA from the bacterium to the plant cell. The genes of the *vir* region encode proteins that are involved in the sensing of the presence of the wounded plant cell and the molecular transduction of this signal to result in the production of T-DNA transfer intermediates. Further, the *vir* region probably also encodes proteins responsible for the passage of the T-DNA intermediates into the plant cell. Near the *vir* loci of nopaline Ti plasmids is another gene involved in the synthesis of *trans*-zeatin (*tzs*). Its sequence is similar to *ipt* and it appears to encode a phosphopentyltransferase. The function of this gene is not clear although its absence from octopine Ti plasmids suggests that its function is not essential and that it may be involved in increasing the host range of bacterium (Beaty *et al.*, 1986).

Hairy Root Formation by *Agrobacterium rhizogenes*

Agrobacterium rhizogenes induces hair root disease in a manner analogous to tumour formation by *Agrobacterium tumefaciens*. Virulent *Agrobacterium rhizogenes* contains a large plasmid, the Ri plasmid, which has a *vir* region which is homologous to the *vir* region of the Ti plasmid and also transfers T-DNA to the plant genome. The agropine type Ri plasmids (e.g. pRiAb) transfer two separate T-DNA regions, T_L and T_R, to the plant genome (Fig. 2.3.) whereas the mannopine (e.g. pRi8196) and cucumopine (e.g. pRi2659) Ri plasmids appear to have a single T-DNA region. The mechanism of transfer of T-DNA from the Ri plasmid to the plant cell appears to be the same as that employed by the Ti plasmid. The T_L region is flanked by 25 bp repeats and to the right of the T-DNA is an overdrive sequence. Compared to the T-DNA of the Ti plasmid, very little is known of the functions encoded by the T-DNA of the Ri plasmids. In a way similar to the octopine T-

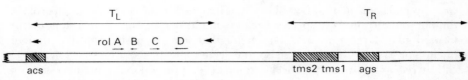

Fig. 2.6   Functional organization of pRiA4 T-DNA. Regions similar to known pTi genes are shown as hatched boxes; acs, agrocinopine synthase; tms, genes homologous to the auxin biosynthetic genes of octopine T_L DNA; ags, agropine synthase. Small arrows indicate the positions of the border repeat sequences, where known.

DNA, the copy number of T_L and T_R T-DNA of the agropine type Ri plasmids is not always the same in transformed tissue and whereas the presence of T_L is essential for the hairy root phenotype, T_R is not, and may be absent altogether. The T_L of the agropine type Ri plasmids has been sequenced and found to contain 18 open reading frames of which ORFs 10, 11, 12 and 15 correspond to the *rolA, B, C* and *D* loci which are important in affecting the virulence on different host plants, although their exact function is not known (Fig. 2.6) (Slightom *et al.*, 1986). Recently is has been shown that ORFs 10, 11 and 12 alone are sufficient to induce hairy roots on tobacco (Cardarelli *et al.*, 1987a; Spena *et al.*, 1987). None of the ORFs has extensive sequence homology with octopine type T-DNA. The T_R region contains an agropine synthase gene as well as genes which are homologous to the auxin biosynthetic genes of octopine T_L DNA, but these are not absolutely required for the maintenance of the hairy root phenotype. In view of the fact that the auxin genes of the Ti plasmid T-DNA are responsible for the tumour phenotype, it may at first appear paradoxical that this is not the case with Ri T-DNA. However it is possible that the gene products of the Ri T_L DNA make the transformed plant cell more responsive to auxins either synthesized by the plant itself or, in the case of the agropine type T-DNA, auxins encoded by T_R. Hence T_R may be seen as an accessory DNA acting to extend the virulence of the agropine-type strains of *Agrobacterium rhizogenes* on different host plants (Cardarelli *et al.*, 1987b).

Mechanism of DNA Transfer from *Agrobacterium* to the Plant Genome

Transfer of DNA from *Agrobacterium* to the plant cell involves a cascade of events requiring the active participation and interaction of both the plant cell and the bacterium. Although recently much light has been shed on the initial steps within the bacteria involved with the transfer, many of the steps, particularly those involved with the passage of the DNA into the plant cell and its integration into the nuclear DNA, remain unknown. The matter is complicated further by variations in both host range and virulence of the bacteria to different plants making it difficult to distinguish between whether a gene function is absolutely necessary for the transfer of DNA to the plant cell or if it is merely involved in the efficiency of tumour formation. This variation arises from the differences between both the *vir* regions and the T-DNAs present on the Ti plasmids of different strains of *Agrobacterium*. Moreover, it must be remembered that tumour formation involves the transfer of DNA from the bacterium to the plant cell as well as the response of the plant cell to the synthesis of plant hormones. Hence, the developmental stage of the plant cell which is to be transformed is critical in tumour formation as tumours can often only be incited on a particular part of the plant, presumably because of the local hormone levels in the infected cell. Therefore any block in tumour formation may result from either the inability of the bacterium to transfer its T-DNA to the plant cell or the plant cell not responding to changes in hormone levels. Bearing these difficulties in mind, the mechanism of tumour induction by *Agrobacterium tumefaciens* can for convenience be divided into a number of steps, and these are discussed below.

Chemotaxis and Plant Cell Conditioning

It is now clear that wounded plant tissue releases several phenolic derivatives into the rhizosphere and one, acetosyringone, acts at low concentrations as a chemi-attractant to the *Agrobacterium* (Ashby *et al.*, 1987). At the same time nopaline-type *Agrobacterium* synthesize and secrete *trans*-zeatin which is thought to 'condition' plant cells to transformation, possibly by inducing cell division. Several bacterial chromosomal loci are involved in the plant–bacterial interaction and mutations in these genes are pleotropic and result in the bacteria being unable to attach to the plant cell wall. The chromosomal genes *chvA* and *chvB* are thought to be involved in synthesizing cell wall components with *chvB* mutants not being able to synthesize β-1,2-glucan (Douglas *et al.*, 1985), whereas *pscA* mutants have an altered cell wall polysaccharide composition (Thomashow *et al.*, 1987) (Fig. 2.7).

Induction of the vir Loci

Expression of the *vir* genes appears to be under the control of at least two regulatory mechanisms: *virA* and *virG* are expressed constitutively at significant levels in the bacteria whereas *virC* and *virD* are expressed at very low basal levels in un-induced bacteria and appear to be controlled by the chromosomal *ros* locus. When the bacteria are exposed to wounded plant cells the expression of *virB, C, D, E* and *G* is induced to high levels. There are several other loci, called *pin* (plant-inducible) loci, which are also induced but do not appear to be required for virulence and may encode functions that act during the bacterial–plant cell interaction (Fig. 2.8) (Stachel and Zambryski 1986a). The induction of the *vir* and *pin* loci is mediated by the phenolic compounds acetosyringone and α-hydroxyacetosyringone which are released by the wounded plant cell (Stachel *et al.*, 1985). The *virA* product is a protein associated with the bacterial membrane which is thought to act as a sensory molecule, sensing the presence of acetosyringone in the rhizosphere and in turn interacting with the *virG* product which induces the transcription of the *vir* and *pin* loci. Sequence analysis of the *virA* and *G* genes reveals a similarity to the bacterial *envZ* and *ompR* genes which are known to be involved in the response of bacteria to changes in osmolarity. This

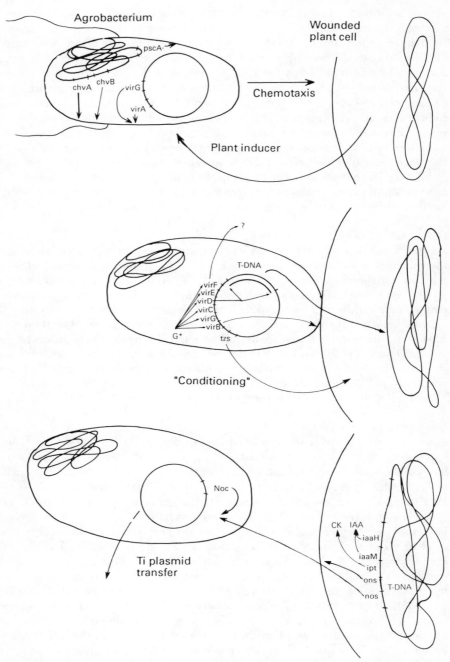

Fig. 2.7   Schematic diagram of the steps involved in the transformation of plant cells by *Agrobacterium tumefaciens*.

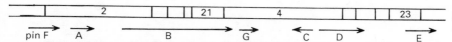

Fig. 2.8 Transcriptional map of the virulence region of pTiA6. Numbers refer to the fragments obtained by digestion with EcoRI.

suggests that sensory transduction in these two cases involves a two component system, a membrane-bound sensor molecule (*virA*) interacting with a mediating molecule (*virG*) which induces the transcription of a variety of genes (the *vir* and *pin* loci) (Ronson *et al.*, 1987).

Expression of the vir *Loci*
Following its interaction with the acetosyringone-activated *virA* product, the *virG* protein induces the transcription of not only itself but also *virB, C, D, E* and *G* (Stachel and Zambryski 1986a). Analysis of the sequences upstream from the *virG*-induced *vir* genes has shown that there are regions of conserved sequences at which the *virG* gene product might bind. High expression of the *virG* locus can result in a supervirulence phenotype (Jin *et al.*, 1987).

The *vir* loci encode a large number of polypeptides some of which are absolutely essential for tumour formation (*virA, B, G* and *D*) whereas others are not (*virC, E, F*) (see Table 2.1). Besides *virA* and *G*, only the functions of two open reading frames of *virD* are known and they encode a site-specific endonuclease activity that

Table 2.1 The genes encoded by the *vir* region of the Ti plasmid

Loci	Coding regions	MW products (Kd)	Required for virulence	Comments
virA	1	91.6	Yes	$\equiv env Z$
virB	9		Yes	
virC	2	25.5 22.7	No	*virC* activity requires both
virD	4	16.2 47.4 21.3 21.3 75.8	Yes	Single-stranded endonuclease
virE	2	7.1 60.5	No	Acts extra-cellularly
virF			No	Acts extra-cellularly
virG	1	29.6	Yes	$\equiv omp R$

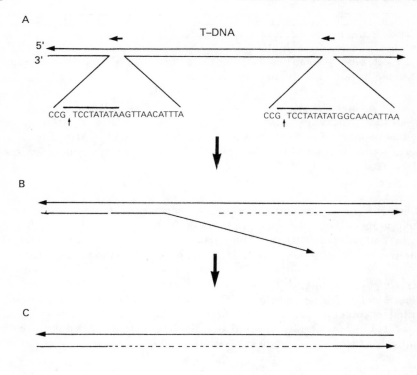

Fig. 2.9   Model for the production of T-strand DNA. (A). *virD* gene product makes site-specific single-strand breaks in the bottom strand of the T-DNA. (B). Unidirectional replication displaces the bottom strand of the T-DNA. (C). The product of replication is double-stranded T-DNA and the T-strand DNA ready for transfer to the plant cell. Small arrows indicate the positions of the border repeat sequences and vertical arrows to site of the single-strand breaks. The thick line represents sequences which are homologous to nopaline T-DNA borders and the broken line represents newly-synthesized DNA. The diagram represents the T_L DNA of pTiA6 (for details see Albright *et al.*, 1987).

is involved in the production of T-DNA transfer intermediates (Yanofsky *et al.*, 1986). Intriguingly, *virE* and *F* appear to act extra-cellularly (Otten *et al.*, 1984).

Production of T-DNA Intermediates
Induction of the *vir* loci results in the appearance of both site- and strand-specific nicks in the bottom strand of the 25 bp border sequences of the T-DNA and the appearance of a single-stranded DNA, the T-strand, that corresponds to the

bottom strand of the T-DNA with its 5′ and 3′ ends mapping to the right and left borders, respectively (Fig. 2.9). The nicks appear approximately 12 h after the bacterial cells have been exposed to acetosyringone and occur between the third or fourth base (± 1 or 2 bases) from the left-hand side of the 25 bp border repeats (Albright *et al.*, 1987; Wang *et al.*, 1987b). There appears to be approximately one copy of the T-strand per induced bacterium (Stachel *et al.*, 1986). Although the number of *vir* loci involved in the individual steps of this process are unknown, it is known that *virD* encodes the site specific endonuclease that acts at the 25 bp repeats (Yanofsky *et al.*, 1986). The T-strand may arise from strand displacement following DNA synthesis. It is thought that the T-strand acts as the T-DNA transfer intermediate. The presence of overdrive in octopine Ti plasmids appears to maximize the levels of T-strand intermediates produced in an infection (Van Haaren *et al.*, 1987). Nopaline Ti plasmids do not appear to have a sequence similar to overdrive but the sequences surrounding the border repeats seem to be important in DNA transfer with the flanking regions of the right and left borders respectively enhancing or attenuating tumour formation (Wang *et al.*, 1987a).

Transfer of DNA to the Plant Cell
The precise mechanism by which DNA is transferred to the plant cell is not known but is thought that the T-strand is transferred in a manner analogous to bacterial conjugation. This process involves one strand of DNA, protected as a DNA-protein complex, being transferred from the donor to the recipient bacteria with its 5′ end leading (Stachel and Zambryski 1986b). Indeed, recently it has been found that plasmid sequences normally utilized in DNA transfer during conjugation, *oriT* and *mob*, can direct the passage of DNA to the plant cell (Buchanan-Wollaston 1987). The observation that genes containing introns can be transferred stably into the plant genome (see Chapter 6) suggests that an RNA transfer intermediate is not employed. A single bacterium can transfer more than one copy of T-DNA to the plant cell during one infection event. During infection the majority of the singly-transformed plant cells are infected by a single bacterium and many of these plant cells contain more than one T-DNA copy integrated into the nucleus suggesting that replication of the T-DNA takes place either within the bacterium or the plant cell prior to stable integration into the plant genome (Depicker *et al.*, 1985).

Stabilization of the T-DNA within the Nucleus
The mechanism by which this takes place is unknown and at present the only way to investigate this is by analysing the T-DNA in the transformed plant cell. It is generally considered that the T-DNA integrates at random into the plant genome. Recently it has been found that the T-DNA in plants transformed with *Agrobacterium tumefaciens* C58 is present predominantly in the form of inverted repeats (Jorgensen *et al.*, 1987). However, the mechanism by which these structures are formed is unknown but could involve replication or ligation of the DNA prior to integration. As shall be seen later (Chapter 5) there is evidence to suggest that when naked DNA is taken up into the plant cell it concatenates prior to integration. Sequencing of the T-DNA inserted into the plant genome has

revealed that the ends of the T-DNA are located close to, or in, the right 25 bp repeat whereas at the left-hand side, the ends may be spread over 100 bp. Comparison of the plant sequence before and after insertion of the T-DNA has shown that complex rearrangements of the plant DNA, presumably during insertion of the T-DNA, can take place, including a duplication of 158 bp of 'target DNA' as well as small deletions and rearrangements (Gheysen *et al.*, 1987). These authors built a model for the mechanism of integration of T-DNA, consistent with our current knowledge, as a four-step process following passage of the T-strand into the plant cell:

(1) A protein at the 5′ (right) end of the T-strand interacts with a nicked sequence in the plant DNA
(2) Local torsional strain on the plant DNA results in the production of a second nick on the opposite strand of the 'target sequence'
(3) The T-strand is ligated to the plant DNA and copied by cellular enzymes
(4) The repair and replication of the staggered nicks in the plant target DNA results in the production of a repeated sequence (of variable length) and of additional rearrangements (additions and/or deletions)

Because there appears to be no firm evidence for the involvement of *Agrobacterium*-encoded polypeptides in this process, it is considered that the plant's normal recombination and repair mechanisms are responsible (Gheysen *et al.*, 1987).

Expression of the T-DNA and Establishment of the Transformed Phenotype
As we have seen, the T-DNA encodes enzymes that presumably start to be expressed after the T-DNA is integrated into the host genome. In the case of the Ti plasmid, these enzymes synthesize auxin and cytokinin which disrupt the hormonal balance of the cell and initiate disorganized growth and the synthesis and secretion of opines. On the other hand, the gene products of the T_L DNA of the Ri plasmid appear to sensitize the transformed cell to the auxins whose synthesis can be directed by the T_R DNA, if it is present, or those that are being synthesized by the plant cell. However, although T-DNA is stably integrated into the plant genome, it might not always be expressed. There is evidence to suggest that methylation of the DNA in transformed tissue can seriously effect its expression, which is an observation that needs to be borne in mind when considering the use of Ti plasmid-based transformation vectors (Hepburn *et al.*, 1983).

The Benefit to the Agrobacterium
The apparent benefit to the *Agrobacterium* of plant transformation is that it provides the bacterium with opines which are used as a source of carbon and nitrogen and can result in the conjugative transfer of the Ti plasmid. The nopaline catabolism genes of the nopaline-type Ti plasmid have been located within the *Noc* region of the Ti plasmid. This region has been found to encode nopaline oxidase, arginase and ornithine cyclodeamidase which converts nopaline to proline (Sans *et al.*, 1987).

References

Albright, L.M., Yanofsky, M.F., Leroux, R., Ma, D. and Nester, E.W. (1987). 'Processing of the T-DNA of *Agrobacterium tumefaciens* generates border nicks and linear, single stranded T-DNA', *J. Bact.* **169**, pp. 1046-1055.

Ashby, AM, Watson, M.D. and Shaw, C.H. (1987). 'A Ti plasmid determined function is responsible for chemotaxis of *Agrobacterium tumefaciens* towards the plant wound product acetosyringone', *FEMS Microbiology Letters* **41**, pp. 189-192.

Beaty, J.S., Powell, G.K., Lica, L., Regier, D.A., MacDonald, E.M.S., Hommes, N.G. and Morris, R.O. (1986). '*Tzs* a nopaline Ti plasmid gene from *Agrobacterium tumefaciens* associated with *trans*-zeatin biosynthesis', *Mol. Gen. Genet.* **203**, pp. 274–280.

Buchanan-Woolaston, V., Passiatore, J.E. and Cannon, F. (1987). 'The *mob* and *oriT* mobilisation functions of a bacterial plasmid promote its transfer to plants', *Nature* **328**, pp. 172–175.

Cardarelli, M., Mariotti, D., Pomoni, M., Spano, L., Capone, I. and Constantino, P. (1987a). '*Agrobacterium rhizogenes* T-DNA genes capable of inducing hairy root phenotype', *Mol. Gen. Genet.* **209**, pp. 475–480.

Cardarelli, M., Spano, L., Mariotti, D., Mauro, M.L., Van Sluys, M.A. and Constantino, P. (1987b). 'The role of auxin in hair root induction', *Mol. Gen. Genet.* **208**, pp. 457–463.

Depicker, A., Herman, L., Jacobs, A., Schell, J. and Van Montagu, M. (1985). 'Frequencies of simultaneous transformation with different T-DNAs and their relevance to the *Agrobacterium*/plant cell interaction', *Mol. Gen. Genet.* **201**, pp. 477–484.

Douglas, C.J., Staneloni, R.J., Rubin, R.A. and Nester, E.W. (1985). 'Identification and genetic analysis of an *Agrobacterium tumefaciens* chromosomal virulence region', *J. Bact.* **161**, pp. 850–860.

Gheysen, G., Van Montagu, M. and Zambryski, P. (1987). 'Integration of *Agrobacterium tumefaciens* transfer DNA (T-DNA) involves rearrangements of target plant sequences', *Proc. Nat. Acad. Sci. USA* **84**, pp. 6169–6173.

Hepburn, A.G., Clarke, R.E., Pearson, L. and White, J. (1983). 'The role of cytosine methylation in the control of nopaline synthase gene expression in a plant tumour', *J. Mol. Appl. Genet.* **2**, pp. 315–329.

Jin, S., Komari, T., Gordon, M. and Nester, E.W. (1987). 'Genes responsible for the supervirulence phenotype of *Agrobacterium tumefaciens* A281', *J. Bact.* **169**, pp. 4417–4425.

Jorgensen, R., Snyder, C. and Jones, J.D.G. (1987). 'T-DNA is organised predominantly in repeat structures in plants transformed with *Agrobacterium tumefaciens* C58 derivatives', *Mol. Gen. Genet.* **207**, pp. 471–477.

Kahl, G. and Schell, J. (1982). *Molecular Biology of Plant Tumors*. London, Academic Press.

Messens, E., Lenaerts, A., Van Montagu, M. and Hedges, R.W. (1985). 'Genetic basis for opine secretion from crown gall cells', *Mol. Gen. Genet.* **199**, pp. 344–348.

Otten, L., DeGreve, H., Leemans, J., Hain, R., Hooykaas, P. and Schell, J. (1984). 'Restoration of virulence of *Vir* region mutants of *Agrobacterium tumefaciens* strain B6S3 by coinfection with normal and mutant *Agrobacterium* strains', *Mol. Gen. Genet.* **195**, pp. 159–163.

Ronson, C.W., Nixon, B.T. and Ausubel, F.M. (1987). 'Conserved domains in bacterial regulatory proteins that respond to environmental stimuli', *Cell* **49**, pp. 579–581.

Salomon, F., Deblaere, R., Leemans, J., Hernalsteens, J-P., Van Montagu, M. and Schell, J. (1984). 'Genetic identification of functions of T_R-DNA transcripts in octopine crown galls', *EMBO J.* **3**, pp. 141–146.

Sans, N., Schröder, G. and Schröder, J. (1987). 'The Noc region of Ti plasmid C58 codes for arginase and ornithine cyclodeamidase', *Eur. J. Biochem.* **167**, 81–87.

Slightom, J.L., Durand-Tardif, M., Jouanin, L. and Tepfer, D. (1986). 'Nucleotide sequence analysis of T_L-DNA of *Agrobacterium rhizogenes* agropine type plasmid', *J. Biol. Chem.* **261**, pp. 108–121.

Spena, A., Schmulling, T., Koncz, C. and Schell, J. (1987). 'Independent and synergistic activity of *rolA*, *B* and *C* loci in stimulating abnormal growth in plants', *EMBO J.* **6**, pp. 3891–3899.

Stachel, S.E. and Zambryski, P. (1986a). '*VirA* and *virG* control plant induced activation of the T-DNA transfer process of *A. tumefaciens*', *Cell* **46**, pp. 325–333.

Stachel, S.E. and Zambryski, P. (1986b). '*Agrobacterium* and the susceptible plant cell. A novel adaption of intracellular recognition and conjugation', *Cell* **47**, pp. 155–157.

Stachel, S.E., Messens, E., Van Montagu, M. and Zambryski, P. (1985). 'Identification of the signal molecules that activate T-DNA transfer in *Agrobacterium tumefaciens*', *Nature* **318**, pp. 624–629.

Stachel, S.E., Timmerman, B. and Zambryski, P. (1986). 'Generation of single stranded T-DNA molecules during the initial stages of T-DNA transfer from *Agrobacterium tumefaciens* to plant cells', *Nature* **322**, pp. 706–712.

Thomashow, M.F., Karlinski, J.E., Marks, J.R. and Hurlbert, R.E. (1987). 'Identification of a new virulence locus in *Agrobacterium tumefaciens* that affects polysaccharide composition and plant cell attachment', *J. Bact.* **169**, pp. 3209–3216.

Van Haaren, Sedee, N.J.A., Schilperoot, R.A. and Hooykaas, P.J.J. (1987). 'Overdrive is a T-region transfer enhancer which stimulates T-strand production in *Agrobacterium tumefaciens*', *Nucleic Acids Res.* **15**, pp. 8983–8997.

Wang, K., Genetello, C., Van Montagu, M. and Zambryski, P. (1987a). 'Sequence context of the T-DNA border repeat element determines its relative activity during T-DNA transfer to plant cells', *Mol. Gen. Genet.* **210**, pp. 338–346.

Wang, K., Stachel, S.C., Timmerman, Van Montagu, M. and Zambryski, P. (1987b). 'Site-specific nick in the T-DNA border sequences following *vir* gene expression in *Agrobacterium*', *Science* **235**, pp. 587–591.

Yanofsky, M.S., Porter, C., Young, L., Albright, L., Gordon, M. and Nester, E. (1986). 'The *virD* operon of *Agrobacterium tumefaciens* encodes a site-specific endonuclease', *Cell* **47**, pp. 471–477.

Chapter 3

Vectors Based on the Ti and Ri Plasmids of Agrobacterium

Features of the Ti and Ri Plasmids Important in Vector Construction

As we have seen in the previous chapter, strains of *Agrobacterium* have evolved an efficient way of genetically engineering plant cells and the mechanism by which they carry this out has several features that make it amenable to exploitation in the construction of plant transformation vectors. These features include the following:

(a) The *onc* genes are not required for the transfer of T-DNA to the plant cell and its integration within the nuclear genome
(b) The *vir* region of the Ti and Ri plasmids functions in *trans*
(c) DNA inserted between the 25 bp border repeats of the T-DNA, whether the borders are natural or synthetic, is transferred to the plant cell
(d) No apparent rearrangements of the DNA located between the T-DNA borders take place during transfer to the plant genome
(e) The foreign DNA integrated into the genome can be stably inherited in a Mendelian manner

Once these aspects of the biological system were apparent it became clear that only the development of both suitable genetic markers and the techniques of plant regeneration from tissue culture were required before *Agrobacterium*-mediated transformation could be used routinely to produce transformed, or transgenic, plants.

The Importance of Genetic Markers in Plant Transformation

One of the most important advances in plant vector construction in particular, and in gene transfer to plant cells in general, was the development of genetic

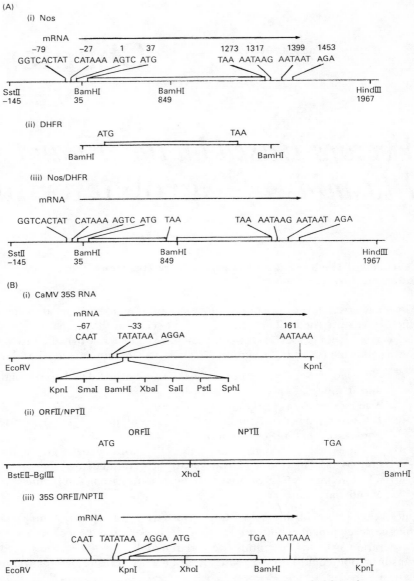

Fig. 3.1 Construction of chimeric genes. Two examples of chimeric gene construction are illustrated. (A). A *nos/DHFR* transcriptional fusion. The *nos* gene was engineered to contain a BamHI site just upstream from the ATG. The DHFR gene contained on a 370-bp BamHI fragment was ligated into the BamHI site to create a transcriptional fusion (Herrera-Estrella *et al.*, 1983a). (B). A CaMV 35S RNA promoter and ORFII NPTII translational fusion. The 35S RNA promoter is contained on a EcoRV–KpnI fragment and just downstream from the transcriptional start site has been engineered a

markers applicable in plant tissue. These bring about in the plant cell a phenotypic change which demonstrates that foreign DNA has entered the cell and that it is not only expressed but also being passed onto progeny. Moreover, the development of dominant selectable markers functional in plant tissue allows the direct selection of transgenic cells by their ability to grow and proliferate under selective conditions. This ensures the recovery of transgenic material which might only be produced at very low frequencies. Such marker or 'reporter' genes can be used to perform a variety of functions and several are available each with their own particular use or advantage. In choosing the appropriate marker gene to be used in the production of transgenic tissue, several characteristics need to be borne in mind. First, it must be decided whether a gene is to be used as a dominant selectable marker, so as to select directly for the growth of transgenic tissue, or whether it will be a screenable marker which can be used to investigate levels of gene activity. Second, the marker gene needs to be small and not contain restriction sites which might interfere with the manipulation of the foreign DNA. Finally, the enzyme encoded by the particular gene must have an activity that is not already present in the plant cell and which can be assayed with a method that is relatively easy to perform using a substrate that is readily (and cheaply!) available. If the assay is to be carried out without disrupting the cell (or tissue) it needs to be unaffected by different cellular environments and the substrate must be able to enter the plant cell (or tissue). In some cases it might be desirable for the enzyme activity to be stable and not readily decay, although this of course is not the case if changing levels of expression need to be assayed, for example during the cell cycle. As yet there are no marker genes that fulfil all of these criteria and generally the choice is made with a particular aim in mind.

Apart from the opine synthases, the marker genes that have been developed are generally gene chimeras where the protein coding sequence is flanked at the 5' end by DNA sequences which can act as a promoter in plant cells and at the 3' end by DNA sequences encoding the addition of a poly(A) tail to the transcribed mRNA. The promoters that have been routinely used are the 35S RNA promoter from Cauliflower Mosaic Virus (CaMV) or the opine synthase promoters, either *ocs* or *nos* from the T-DNA itself. The 35S RNA promoter directs the synthesis of the 35S RNA which is thought to be the replicative intermediate of CaMV (Chapter 4). These promoters have been characterized in great detail and are generally considered to be constitutive in their mode of expression. This is necessary if the marker is to provide dominant selection because selection might be applied to the

---

multiple cloning site. The ORFII/NPTII translational fusion consists of ORF II of CaMV fused, at the XhoI site, in frame with the NPT gene from Tn 903. The fusion is on a BstEII–BamHI fragment and cloned into the promoter construct by cleaving with BstEII and BamHI, blunt-ending the BstEII site and cloning into the SmaI and BamHI cut vector (Pietrzak *et al.*, 1986). Sequences important in the transcription and translation of the constructs are as shown where applicable with their positions numbered with respect to the start of transcription; open reading frames are represented by open boxes. (Not to scale.)

cells at different levels of differentiation or development (i.e. as protoplasts, callus or whole plants). The promoter can be fused to the marker gene outside of the protein coding region to produce a transcriptional fusion or inside the protein coding region to produce an in-frame translational fusion. In the latter case it is important that the addition of amino acids to the amino terminal of the protein does not seriously affect its enzymatic activity. With these gene chimeras it is of course important to ensure that they are constructed so that there are no frame-shifts or spurious translation stop signals to interfere with translation of the resultant mRNA and that there are no initiation codons inserted upstream from the authentic ATG of the protein coding region (Fig. 3.1). Generally the *nos* or *ocs* poly(A) addition sequence is added downstream from the protein coding region.

Genetic Markers for Use in Plant Cells

Opine Synthases (nos *or* ocs)
As we have seen in the previous chapter the opine synthase genes are uniquely encoded by the T-DNA and being constitutively expressed are useful biochemical markers. Moreover, being located near the right border of the T-DNA they are invariably transferred to the plant cells and so have been used extensively in establishing whether a tissue is transformed. The assay involves the electrophoretic separation of amino acids extracted from tissue which has been incubated overnight in arginine (α-ketoglutarate can also be added). Those amino acids, such as arginine and nopaline (or octopine) containing a guanidinium can be visualized under UV light after staining the electrophoretogram with phenanthrenequinone (Fig. 3.2) (Otten and Schilperoot 1978). Although the assay is quite straightforward it is not easily quantified and so these markers are not those of choice for investigating levels of gene expression.

β-Galactosidase (lacZ)
The *lacZ* gene of *E. coli* has been used extensively as a genetic marker in a variety of cells because following its induction with isopropyl-thiogalactoside (IPTG) it reacts with 5-bromo-4-chloro-3-indoly-β-D-galactopyranoside (X-gal), which is a chromogenic substrate, to produce a blue colour. The *lacZ* has been inserted into sunflower and tobacco cells as a translational fusion with the *nos* promoter, the *nos* N-terminal region and the poly(A) addition site. β-galactosidase activity can be visualized following electrophoresis on non-denaturing polyacrylamide gels by staining with 4-methylumbelliferyl-β-D-galactosylpyranoside from which β-galactosidase releases a fluorescent product that can be visualized (Helmer *et al.*, 1984). The marker is useful because it can tolerate additional amino acids at its amino terminus although it is impractical in many plant systems because of endogenous β-galactosidase activity.

Neomycin Phosphotransferase (*NPTII*) (neo)
Neomycin phosphotransferase (NPTII), encoded by the bacterial transposon Tn 5, confers resistance to aminoglycoside antibiotics and has become an

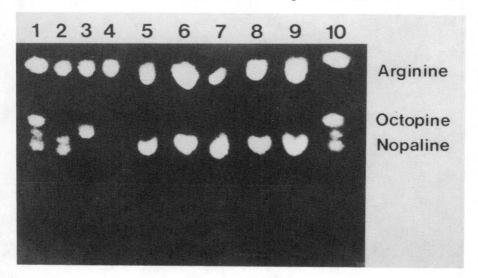

Fig. 3.2   Opine assays. Electrophoretogram demonstrating the presence of *nos* and *ocs* activity in several putative transformants. Samples of tobacco were subjected to paper electrophoresis and following staining visualized by UV light. (Photograph by Rod Scott.)

important marker for use in plant cells largely because selection for resistance can be applied to isolated cells, callus, tissue explants and whole plants (Herrera-Estrella *et al.*, 1983a; DeBlock *et al.*, 1984). Moreover, the enzyme can tolerate extensive modifications at its amino terminal without the destruction of its enzyme activity. There is an enzyme assay available for NPTII which involves the electrophoretic separation of the enzyme on non-denaturing polyacrylamide gels (Reiss *et al.*, 1984). In order to carry this out a cell extract is electrophoresed and the gel equilibrated against the reaction buffer. An agarose gel containing $[\gamma\text{-}^{32}P]ATP$ and kanamycin sulphate is cast onto the non-denaturing gel. The enzyme activity produces phosphorylated kanamycin which is transferred to paper by blotting and visualized by autoradiography (Fig. 3.3). This method has the advantage of being able to detect differences in the size of the protein product of different gene fusions because of their changed electrophoretic mobility but has the disadvantage of using very high levels of radioactivity. In addition, the kanamycin resistance gene of Tn 903 can also be used as a selectable marker, but it appears that the NPT type II gel assay does not work for the type of enzyme encoded by Tn 903 (Pietrzak *et al.*, 1986).

Chloramphenicol Acetyl Transferase (CAT) (cat)
Derived from the bacterial transposon Tn9, chloramphenicol acetyltransferase (CAT) is a useful genetic marker because of its small size, its ability to tolerate small modifications at its amino terminal (An 1986) and the ease of assaying its

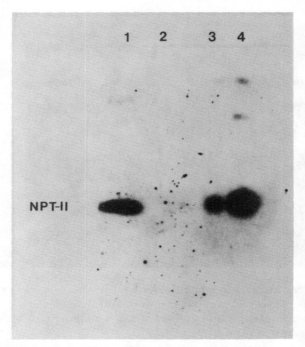

Fig. 3.3 NPTII assay. NPTII activity visualized *in situ* in a non-denaturing polyacrylamide gel by auroradiography. (1) *E. coli* containing the NPTII construct; (2) Non-transformed plant; (3) and (4) extracts of two transformed plants containing the NPTII construct. (Photograph by Rod Scott.)

enzyme activity. The assay for CAT involves the separation in ethylacetate of acetylated chloramphenicols, the products of the enzyme reaction, from non-acetylated chloramphenicol followed by thin-layer chromatography using chloroform and methanol as solvents. ^{14}C-chloramphenicol is used as a substrate and the three forms of the acetylated chloramphenicol, two monoacetates (1-acetylchloramphenicol and 3-acetylchloramphenicol) and one diacetate (1,3-diacetylchloramphenicol), which have different mobilities can be visualized by autoradiography (Fig. 3.4) (Herrera-Estrella *et al.*, 1983a, 1983b). Although CAT has been used as a dominant selectable marker, many plant systems are not sensitive to chloramphenicol and it has not proved to be as versatile as NPTII when used as a selectable marker (DeBlock *et al.*, 1984).

Streptomycin Phosphotransferase (SPT)
The streptomycin resistance gene, encoding streptomycin phosphotransferase (SPT) and derived from Tn 5, has been engineered into tobacco and confers resistance to streptomycin at levels of 1 mg/ml. Streptomycin does not kill plant cells but sensitive cells become bleached and suffer retarded growth whereas resistant ones remain green and continue to proliferate (Jones *et al.*, 1987).

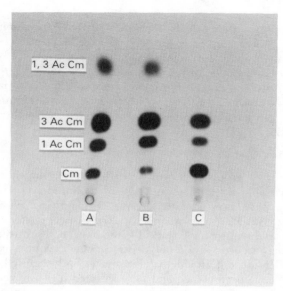

Fig. 3.4 CAT assays. The radiolabelled acetylated derivatives of chloramphenicol are separated by chromatography and visualized by autoradiography. Extracts tested contain a 35S RNA promoter–CAT fusion. (A) transgenic tobacco callus; (B) transgenic tobacco leaf tissue; (C) control bacterial extract. (Photograph by Dianne Firby.)

Dihydrofolate reductase (DHFR) (dhfr)
Dihydrofolate reductase (DHFR), which is involved in the conversion of dihydrofolate to tetrahydrofolate, is inhibited by methotrexate which results in impaired RNA and DNA synthesis. A gene has been derived from *E. coli* that has been found to be insensitive to methotrexate and in an appropriate construct confers resistance to methotrexate in plant cells (Herrera-Estrella *et al.*, 1983a). Similar results have been obtained in transforming oilseed rape using a methotrexate resistant DHFR gene from mice (Pua *et al.*, 1987). The DHFR gene is relatively small and as shall be seen in Chapter 4 this has allowed it to be used in conjunction with a vector based on a viral genome to confer methotrexate resistance to turnip.

Hygromycin Phosphotransferase (aphIV)
Hygromycin B is another aminoglycoside antibiotic which disrupts ribosome function in a variety of cell types. Resistance in *E. coli* is encoded by a plasmid-borne hygromycin phosphotransferase gene which, when transferred to tobacco cells, confers resistance to levels of hygromycin of up to 50 mg/l (Waldron *et al.*, 1985).

Bleomycin

Bleomycin is a polypeptide which interacts with DNA producing single- and double-stranded breaks. A gene encoded by the bacterial transposon Tn 5 produces a 126 amino acid polypeptide which confers resistance by an unknown mechanism in a number of organisms. Cells of *Nicotiana plumbaginifolia* and tomato appear to be sensitive to the antibiotic and resistance has been introduced into tobacco with transformed calli being selected by resistance to bleomycin at 10 μg/ml. Transcription of the gene was demonstrated by Northern blot analysis (Hille *et al.*, 1986).

Luciferase

The luciferases are a group of enzymes which catalyse the chemical reactions which are responsible for light production in bioluminescent organisms. The luciferase genes have been isolated from both firefly (*Photinus pyralis*) and luminescent bacteria (*Vibrio harveyii* and *V. fischeri*). Insect luciferases are encoded by a single gene and require luciferin, ATP and oxygen as substrates whereas the bacterial *lux* operon encodes two polypeptides which require oxygen, reduced flavin mononucleotide and an aldehyde substrate. Both the firefly and bacterial luciferase genes have been cloned and once inserted into a novel bacterial host under the control of an appropriate promoter sequence can produce bioluminescence in the presence of the appropriate substrates. Such genetic markers are of great interest because they are sensitive, do not involve radioactivity in their assays and can be assayed by non-destructive methods. In one report (Ow *et al.*, 1986), the firefly luciferase gene was linked by a transcriptional fusion to the 35S RNA promoter of CaMV and the *nos* poly(A) addition site. This construct was introduced into carrot protoplasts by electroporation (see Chapter 5) and into tobacco plants by the use of Ti plasmid-mediated transformation. Transient expression of luciferase in the carrot cells, in the presence of ATP and luciferin, could be assayed 24 h after electroporation by using a luminometer. Following watering the transgenic plants containing the same construct with a solution of luciferin, luciferase activity could be visualized by placing the intact plant next to photographic film. However, there were large differences in activity in different tissues. The reason for this variation could be due to a variety of factors including the accessibility of the luciferase to the luciferin, compartmentalization of either the enzyme or the substrate or differential absorption of the emitted light. On the other hand, enzyme activity could also be measured by isolating cells from the transgenic plants and incubating them in the presence of luciferin in sterile plastic dishes and placing the dishes next to X-ray film. This then provides a simple method for assaying the gene expression in a non-destructive manner. Another report describes the transfer of both the *luxA* and *luxB* subunits of *V. harveyii* to plants (Koncz *et al.*, 1987). In this case the genes were placed under the control of the dual promoter sequences of genes 1′ and 2′ from octopine T_R T-DNA (see Fig. 2.4, p. 13). The genes introduced by electroporation were transcribed to produce an active heterodimeric enzyme. Apart from being the first report of an assembly of an active foreign enzyme from two subunits in plants, it also proved to be a very sensitive marker with the activity in cell extracts being measured by a luminometer. The difficulty with the genes from *V. harveyii* is that both *luxA* and *luxB* are required for the assembly of active luciferase; however there has been a preliminary report that the two genes have been linked in a translational unit to produce an active luciferase (Schell 1987).

Glucuronidase (GUS) (uidA)

β-Glucuronidase (GUS), derived from *E. coli*, is a hydrolase that cleaves a wide variety of β-glucuronides. The enzyme is very stable, active in a variety of cellular environments and there is little or no β-glucuronidase activity in plants. Moreover, the enzyme can tolerate large amino-terminal additions. GUS is a useful marker largely because it is able to cleave a variety of commercially available substrates for a variety of spectrophotometric, fluorometric and histochemical assays. These can be particularly useful in investigating the tissue-specific expression of a particular gene construct (Jefferson *et al.*, 1987). However, care needs to be taken to ensure that the tissue to be assayed is free of any contaminating bacteria that also contain GUS activity. In addition, the enzymes used in the preparation of protoplasts appear to contain GUS activity and this necessitates their removal prior to carrying out transient expression assays.

Threonine dehydratase (IVL1)

One of the most powerful techniques in genetics is the correction of auxotrophic mutants by complementation with the wild type allele of the mutated gene(s). An example of this in plant cells has been provided by the complementation of a *Nicotiana plumbaginifolia* mutant which requires isoleucine by a gene derived from yeast (Colau *et al.*, 1987). In this case the mutant was found to be deficient in threonine dehydratase which is involved in the first step in the synthesis of isoleucine. The yeast gene encoding this enzyme, *IVLI*, was inserted between the *nos* promoter and *ocs* poly(A) addition site and transferred using a Ti plasmid binary vector (see later) into mutant cells. Resulting calli and shoots could grow in the media lacking isoleucine and contained threonine dehydratase enzyme activity.

Metallothionein II (CHMTII)

Metallothioneins (MT), derived from mammalian cells, are a group of low molecular weight polypeptides which selectively bind the heavy metals Cu, Zn and Cd. Increased levels of expression of MT in animal cells results in the increased tolerance of the cells to heavy metals. Plants contain phytochelatins which seem to have a similar role to mammalian MT although plants are sensitive to Cd. A cDNA clone of the metallothionein from Chinese Hamster has been cloned into a viral vector (Chapter 4) and found to impart to the systematically infected tissue increased Cd binding capabilities so that the excised leaves remained 'healthy' following exposure to Cd in solution (Lefebvre *et al.*, 1987).

Herbicide tolerance

As will be discussed in more detail in Chapter 7, several groups have an interest in engineering herbicide tolerance into crop plants and the genes involved can also be used as dominant selectable markers. To date, two examples of using tolerance to herbicides for selection purposes have been reported, one using tolerance to glyphosate (Shah *et al.*, 1986) the other to phosphinothricin (PPT) (DeBlock *et al.*, 1987).

Glyphosate inhibits enolpyruvylshikimate-3-phosphate (EPSP) which, being involved in the shikimate pathway, is important in the synthesis of aromatic amino acids. Tolerance to glyphosate can be obtained by over-production of EPSP and this has been obtained in transgenic tissue by fusing the EPSP gene to the 35S RNA promoter and the *nos* poly(A) addition site. Selection could be applied both at the callus level at 0.5 mM glyphosate and at the level of regenerated plants which are tolerant to 2–4 times the amounts of glyphosate required to kill wild type plants by spraying.

Glutamate synthetase is important in nitrogen metabolism within plants, being the only enzyme to detoxify ammonia. It is inhibited by PPT which is an analogue of L-glutamic acid. Plants are sensitive to both PPT and bialaphos which is a tripeptide antibiotic consisting of PPT and two analine residues produced by *Streptomyces hygroscopicus*. The bialaphos resistance gene (*bar*) from *S. hygroscopicus* encodes phosphothricin acetyltransferase (PAT) which acetylates a free amino group of PPT and detoxifies it. This gene engineered for expression in plant cells has been placed downstream of the 35S RNA promoter and fused to the poly(A) addition site of gene 7 of the octopine T-DNA and found to provide resistance of up to 50 mg/l of PPT in calli and whole tobacco, tomato and potato plants.

Potential Markers for Use in Plant Cells

It is likely that the above list of markers will continue to grow to include plant-specific genes encoding unique but easily assayable enzyme activities which might be used as genetic markers as well as genes that can act as negative markers (i.e. those that are lethal to the plant cells or interfere with their development when expressed so allowing the isolation of regulatory mutants). A possible example of the latter type of marker is provided by the *iaaM* gene from *Pseudomonas savastanoi*. It has been found that this gene, when linked to the gene 1 promoter of octopine T_R DNA and transferred to the tobacco genome, induces the growth of callus but appears to inhibit shoot formation (Inze *et al.*, 1987). This system could be used to study gene inactivation by subjecting callus or isolated cells that contain the gene 1 promoter–*iaaM* construct to mutagenesis and selecting those clones that are able to form shoots.

Complementation has proved to be an extremely powerful technique in both isolating genes and studying their importance in biochemical pathways. The results involving complementation of a mutant plant cell with the yeast threonine dehydratase indicate that complementation is feasible with plant cells. Complementation has also been demonstrated in a regenerated transformed plant (Meyer *et al.*, 1987). In this study the complementation of a mutant *Petunia hybrida* which accumulates the flower pigment precursor dihydrokaempferol was attempted. Generally, in plants the enzyme dihydroflavonol 4-reductase (DFR) converts hydrokaempferol to the immediate precursors of anthocyanins which lead to flower pigmentation. However, the *Petunia* DFR cannot accept dihydrokaempferol as a substrate and there is a mutant line of *Petunia* which is unable to synthesize flower pigments by the normal pathway, accumulates dihydrokaempferol and has flowers that are pale pink in colour. Naked DNA uptake (see Chapter 5) was used to insert the maize dihydroquercetin 4-reductase (DQR) gene fused to the 35S RNA promoter and the *ocs* poly (A) addition site into the mutant *Petunia* line. DQR is involved with pigment formation in the aleurone layer of maize kernels. The flowers of the transgenic plants obtained were coloured brick red and accumulated no dihydrokaempferol indicating that the maize DQR was able to utilize the dihydrokaempferol that accumulated in the mutant *Petunia* and convert it to anthocyanins thus creating a new biochemical pathway.

Another potential example of complementation could be with nitrate reductase now that a cDNA encoding this has been isolated and characterized (Crawford *et al.*, 1986) and cell lines deficient in nitrate reductase have been obtained which can be regenerated into whole plants (Müller 1983).

Strategies for Constructing Vectors Based in the Ti or Ri Plasmid

Possibly every laboratory that has worked on the molecular biology of *Agrobacterium* has developed its own vector system for plant transformation based on either the Ti or Ri plasmid and for the non-expert there is a confusing array of vector molecules which are continually being modified and improved. However, amidst this variety the individual vector can generally be classified according to whether the foreign DNA which is to be transferred to the plant cell is physically linked to the *vir* region of the Ti or Ri plasmid, or whether it is separate. The former are known as co-integrative or *cis* vectors, whereas the latter are referred to as binary or *trans* vectors. Although the vector plasmids may differ from each other, all rely on the *Agrobacterium* in which it is resident being able to carry out all of the steps in the process of DNA transfer to the plant cell.

Co-integrative vectors are those based on a wild type Ti or Ri plasmid from which portions of the T-DNA have been removed, or replaced by a novel sequence of DNA. Often, with the Ti plasmid, the region that has been removed encodes the *onc* functions and this allows regeneration of normal non-tumourous plants using conventional procedures of tissue culture. These vectors are often called 'disarmed vectors'. Because they are based on the Ti or Ri plasmid, they are stable within the *Agrobacterium* and because they retain the *vir* region, have all of the apparatus necessary to transfer sequences located between the border repeats to the plant cell. In order to insert foreign DNA between the T-DNA borders it can be cloned on an intermediate vector that can be manipulated in *E. coli* and which has a region of homology with the sequence between the border repeats of the co-integrative vector. The plasmid is transferred from *E. coli* to the *Agrobacterium* which contains the co-integrative vector by conjugation. The intermediate vector plasmids are not normally transferred by conjugation and require the presence of

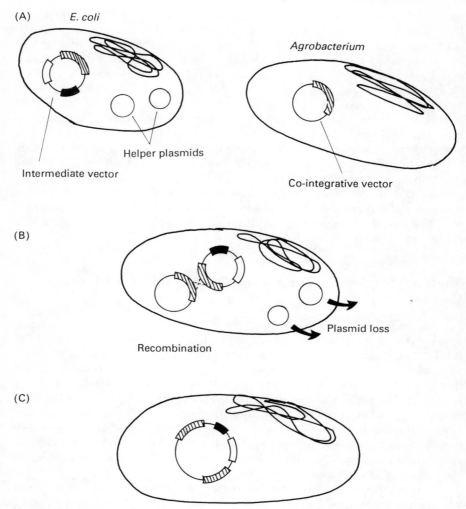

Fig. 3.5 Recombinational rescue. (A). *E. coli* containing helper plasmids and the intermediate vector containing selectable marker (■) foreign DNA (□) and region of homology with the Ti/Ri plasmid (▨). *Agrobacterium* containing a Ti/Ri co-integrative vector plasmid with region of homology with the intermediate vector (▨). The *Agrobacterium* contains a chromosomally-encoded rifampicin resistance gene. (B). Following conjugation, the *Agrobacterium* contains all four plasmids but those from the *E. coli* are lost except where recombination takes place between the region of homology contained on the intermediate and the Ti/Ri co-integrative vector. (C). Transconjugants containing the intermediate vector rescued by recombination with the Ti/Ri co-integrative vector plasmid selected for by resistance to rifampicin and the marker carried by the intermediate vector. (Not to scale.)

helper plasmids, such as pGJ28 and R64*drd*11, which provide ColE1 helper functions allowing conjugative transfer (Van Haute *et al.*, 1983). Once inside the *Agrobacterium*, the plasmids which lack suitable origins of replication are lost; however, provided that there is a region of homology between the intermediate vector and the co-integrative vector recombination can take place between the two so that the former can be rescued and retained in the *Agrobacterium*. Exconjugants which have undergone recombination can be selected for by an appropriate genetic marker such as an antibiotic resistance (Fig. 3.5).

Binary vectors are based on plasmids which can replicate both in *E. coli* and strains of *Agrobacterium* and contain the T-DNA border sequences flanking multiple cloning sites as well as markers that allow direct selection of the transformed plant cell. These vectors allow manipulation in *E. coli* followed by transfer to *Agrobacterium* by conjugation in the presence of helper plasmids (Hoekema *et al.*, 1983). Binary vectors can function in both *A. tumefaciens* and *A. rhizogenes* (Simpson *et al.*, 1986). The binary vector replicates as an independent plasmid in *Agrobacterium* and the transfer of the foreign DNA into the plant cell is mediated by the *vir* region of the resident Ti or Ri plasmid acting in *trans*. In *A. tumefaciens*, the recipient bacterium strain (usually LBA4404) (Hoekema *et al.*, 1983) contains a Ti plasmid from which the T-DNA and the 25 bp border repeat has been removed. Some binary vectors are not stable in *Agrobacterium* and require growth under selective conditions to be maintained. While this may result in the reduction of the number of bacteria containing the vector during inoculation of plant tissue (see later) it may be considered as providing a level of biological containment to recombinant molecules. Because the T-DNA can be transferred independently, plants containing the T-DNA of the binary vector alone can be obtained provided that the correct selection is applied to the transgenic tissue (Fillatti *et al.*, 1987; Trulson *et al.*, 1987).

Using a Vector Based on the Ti or Ri Plasmid

In choosing a vector with which to carry out plant transformation, several factors, not least the availability of the vector, need to be borne in mind. These are now discussed.

Experimental Aims
Experimental aims can be myriad, including promoter analysis, *in vivo* mutagenesis, modification of a plant phenotype and gene tagging. Many vectors have been designed with a particular experiment in mind, hence it is important to select a vector carefully for the task required of it. For example, to assay gene (or promoter) activity it is important to have an assayable marker which is independent from the experimental construct, to ensure that the inserted DNA is functional. A dominant selectable marker can be used to ensure detection of rare transformation events; to be certain that all of the construction has been integrated into the plant genome and is being expressed, two selectable markers located within each border sequence might be used with selection being applied

for both. It is of course important that the dominant marker chosen for the experiment works effectively in the plant species used.

Manipulation of Foreign DNA
Should the DNA sequence of interest have convenient restriction sites it might be relatively straightforward to clone it into a binary vector (or intermediate vector) containing compatible restriction sites. If, on the other hand, this is not so, homologous recombination may be used with co-integrative vectors to insert the sequence of choice within the T-DNA border sequences.

Ease of Transformation of the Target Plant
Generally, the method of transformation adopted depends on the ease of plant regeneration for a particular plant species. In several species particularly amongst the *Solanaceae*, it is possible to obtain transformants readily following explant inoculation (see later) and plant regeneration. Where regeneration may prove more difficult, *A. rhizogenes* can be a useful alternative, utilizing the ability of plants to be regenerated from cultures of 'hairy roots'. Nevertheless, it is becoming increasingly clear that the virulence exhibited by different strains of *Agrobacterium* can vary with the plant host. This is a function of the ability of the *Agrobacterium* to transfer its T-DNA to the plant host and, where tumour induction is monitored, the ability of the plant cell to respond to aberrant levels of hormones. It is probable that host range variation is due to several factors. Recently it has been shown that a binary vector in a C58 *Agrobacterium* background enhanced the rate of shoot proliferation from explants of *Populus*. This could have been due to the genes specifying hormone synthesis present in the T-DNA or the *tzs* gene of the *vir* region (Fillatti *et al.*, 1987). Moreover, a 'supervirulent' Ti plasmid (pTiBo542) in the *Agrobacterium* strain A281 has been found to induce large rapidly growing tumours on a wide range of host plants (Komari *et al.*, 1986). When the plasmid is used as a *vir* helper to a binary vector it results in increasing the efficiency of transformation (An 1985; An *et al.*, 1985). Thus when attempting to transform a new plant species it might be preferable to use a binary vector in a variety of *Agrobacterium* strains containing different *vir* functions to assess which combination is the most efficient in obtaining transformation.

Transformation Vectors based on the Ti and Ri Plasmids

While it is beyond the scope of this book to describe in detail each of the vectors based on the Ti or Ri plasmid that have been developed, a description will be given of some of the vectors that have been used routinely.

pGV3850
pGV3850 is based on a C58 nopaline Ti plasmid from which the *onc* genes have been removed and replaced by pBR322 (Zambryski *et al.*, 1983). Any foreign DNA that has been cloned in an intermediate vector based on pBR322 can be introduced into pGV3850 by conjugation in the presence of helper plasmids

followed by recombinational rescue. Exconjugants are selected for by the presence of antibiotic resistance genes present on the intermediate vector (streptomycin/spectinomycin or kanamycin) and resistance to rifampicin which is encoded by the chromosome of the *Agrobacterium* (Fig. 3.6A).

The SEV System

The split-end vector (SEV) system is another co-integrative vector system based on a derivative of pTiB6S3 from which the *onc* genes, the right border of the T_L DNA and the T_R DNA have been removed. The left border sequence is retained as is a small portion of the T-DNA known as the limited homology region (LIH) Foreign DNA is cloned into an intermediate vector, based on pBR322, which contains selectable markers for both bacteria and plant cells flanked by the right border sequence and the LIH. The intermediate vector is introduced into the *Agrobacterium* containing the co-integrative vector by conjugation. Recombination between the LIH on the two plasmids results in the foreign DNA being introduced into the co-integrative vector between the two border sequences, the right and left borders being provided by the intermediate vector and the co-integrative vector, respectively (Fraley *et al.*, 1985). Once again, exconjugants are selected for by the use of appropriate antibiotics (Fig. 3.6B).

Bin19

Bin19 is based on the broad host range plasmid, pRK252, and is a binary vector which can replicate in both *E. coli* and *Agrobacterium* (Bevan 1984). This allows manipulation of the plasmid by standard DNA technology followed by the introduction into *Agrobacterium* by conjugation. The vector contains two kanamycin resistance genes, one functional in bacterial cells, the other in plant cells. A multiple cloning site allows for the insertion of foreign DNA within the α-complementary region of β-galactosidase. This allows selection of recombinant clones in *E. coli* as being *lac⁻* in the presence of X-gal and IPTG (Fig. 3.7).

AR1193

This is a co-integrative vector derived from the Ri plasmid pRi1193 (Stougaard *et al.*, 1987). It was constructed by sub-cloning two fragments of the T_L DNA and placing pBR322 between them. Homologous recombination between the T_L DNA sequences on the intermediate vector and the wild type pRi1193 results in the integration of the pBR322 sequences in the T_L DNA. Integration of the pBR322 between the two segments of T_L DNA does not disrupt the genes that are encoded by them (Fig. 3.8A). This vector is analogous to pGV3850 and foreign sequences can be introduced by the same method of conjugation followed by recombination.

A. rhizogenes *Intermediate Vectors*

These are vectors based on *E. coli* plasmids that contain an internal sequence of the T_L DNA of *A. rhizogenes*. Recombination between the intermediate vector and the resident Ri plasmid serves to introduce the foreign DNA sequence between the T-DNA border sequences (Fig. 3.8B) (Stougaard Jensen *et al.*, 1986).

(A)

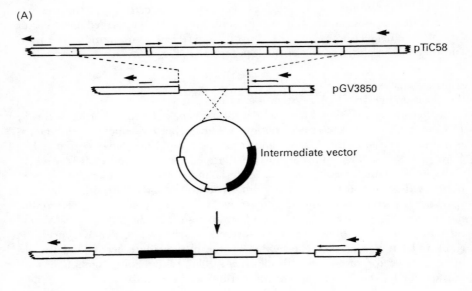

(B)

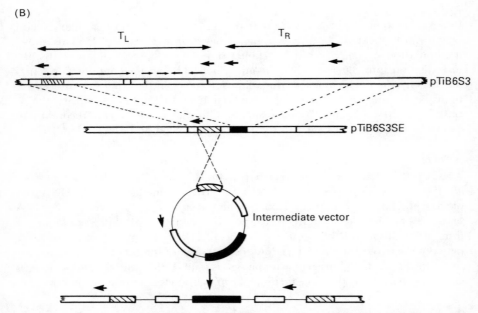

Fig. 3.6 Co-integrative vectors. (A). pGV3850, based on pTiC58 has the *onc* genes replaced by pBR322. The intermediate vector, based on pBR322, containing a selectable marker (■) and foreign DNA (□) is not maintained within the *Agrobacterium* unless recombination has taken place between the homologous pBR322 sequences. (B). pTiB6S3SE is a deleted octopine Ti

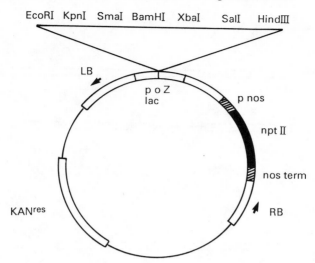

Fig. 3.7 Binary vectors. Bin19 contains a wide host range origin of replication and a kanamycin resistance gene for selection in bacteria. Right and left border repeats flank a kanamycin resistance gene flanked by the *nos* promoter and poly (A) addition site for the selection of transgenic plant tissue and a polylinker cloning site upstream from the *lacZ* gene allowing for selection of clones containing inserts at this site in *E. coli* using a chromogenic substrate. Arrows indicate the position of the border repeat sequences; left border (LB) right border (RB).

---

plasmid from which T_R and the majority of T_L has been replaced by a kanamycin resistance gene. A region of homology (▨) and the left border repeat remains. The intermediate vector contains the right border sequence and *nos* (□, left), a selectable marker (■) sequence, a region for insertion of foreign DNA as well as a chimeric kanamycin resistance gene for selection of transgenic plant tissue (□, right) and a region of homology with pTIB6S3 (▨). Recombination between the region of homology of the intermediate vector and pTiB6S3SE results in the foreign DNA being positioned between the two border repeats on the Ti plasmid. Large arrows indicate the position of the border repeat sequences, small arrows and lines the transcriptional units of the T-DNA and the single line pBR322. (Not to scale.)

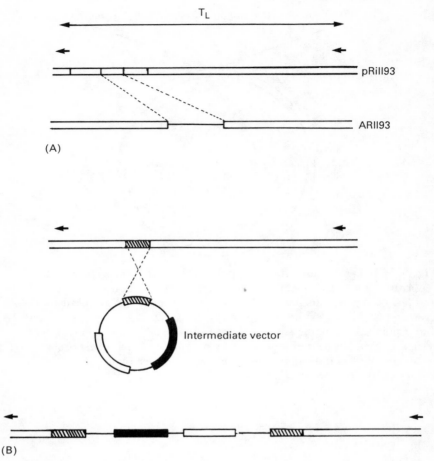

Fig. 3.8 *A rhizogenes* vectors. (A). AR1193 based on the Ri plasmid, pRil193, contains pBR322 inserted towards the left hand side of the T_L DNA to the right of the left border sequence. Recombination results in the intermediate vector being inserted between the border repeats in a similar manner to that seen with pGV3850. (B). An intermediate vector containing a segment of the Ri plasmid (▨), a foreign gene sequence (▢). A selectable marker gene (■) can be introduced into the *Agrobacterium* by conjugation and recombinational rescue between the regions of homology can rescue the foreign sequences of DNA between the border repeats of the T_L DNA. (Not to scale.)

Methods of Plant Genetic Transformation using *Agrobacterium*

Currently there are a variety of methods that can be used to produce transgenic plants using Ti or Ri plasmid-derived vectors. For the Ti plasmid vectors from which the *onc* functions have been removed, three methods have been developed: co-cultivation of protoplasts with *Agrobacterium* followed by callus formation and plant regeneration, explant inoculation with *Agrobacterium* followed by plant regeneration and the most recently developed, *Agrobacterium*-mediated transformation of germinating seeds. For Ri plasmid vectors the situation is somewhat different because these vectors generally contain an unmodified T_L and T_R region and because the T-DNA can be transferred independently, transformants can contain either of the T-DNAs (or both); if a binary vector is used the binary T-DNA can be present alone or with the T_L or T_R DNA, or both. As we have seen, the genes of the T_L region do not encode *onc* functions as such but encode proteins which sensitize the plant cell to plant hormones. This then allows regeneration of plants from transformed hairy root cultures by manipulation of the hormone conditions.

Co-cultivation

Co-cultivation involves isolating protoplasts from sterile leaf tissue and incubating them in culture medium for 2–3 days so that, although they do not divide, they begin to initiate cell wall formation. The cells are then incubated with a fresh culture of *Agrobacterium* for between 1–2 days. Following the co-cultivation, the protoplasts are collected by centrifugation and cultured further in the presence of antibiotics to remove any remaining bacteria. Initially carbenicillin, vancomycin and streptomycin were used for this but more recently Claforan (or cefotaxime) has been adopted as an effective antibiotic. The protoplasts can be cultured to produce callus which can in turn be induced to form shoots of which a proportion will be transformed (Marton *et al.*, 1979). If the DNA contains a selectable marker, transformed tissue can be selected for at the callus or shoot induction stage. The advantage of this technique is that a large number of transformants can be obtained and by carefully controlling the ratio of bacteria to plant cells, the number of T-DNA inserts in the transgenic cells can be controlled.

The co-cultivation technique has been applied to cells derived from suspension culture (An 1985) and extended further for use in transforming a pre-embryogenic suspension culture of carrot (Scott and Draper 1987). This is important because very few economically important plant species can be regenerated from protoplast-derived tissues. On the other hand, regeneration can be achieved with many crop plants from suspension cells by both organogenesis and somatic embryogenesis.

Explant Inoculation

Explant inoculation is possibly the most convenient way of producing transgenic material but depends on two factors: first, the wounded cells must be able to interact with the *Agrobacterium* and second, the cells that have been transformed must be able to divide and regenerate into plants. A wide variety of tissue explants can be used in this sort of experiment: leaves, stems, hypocotyls, roots and tubers. However, it needs to be borne in mind that explants may contain quiescent meristems within the tissue that are not accessible to the *Agrobacterium* but nevertheless are very efficient in producing shoots. Leaf inoculation has proved to be useful for producing transgenic tissue from a variety of plant species including tobacco, tomato, *Petunia* and *Populus*. Devised independently in a variety of laboratories the method involves cutting sterile leaf discs from plants and placing them on agar media. A fresh culture of *Agrobacterium* is applied to the discs and incubated overnight. The discs are then removed to the media containing Claforan which induces callus formation at the wounded periphery of the disc and the callus is subsequently removed and placed on media to induce shoot formation. In the absence of selection, up to 20% of the shoots may be transformed. Selection can be applied at either the callus or shoot-induction stage (Rogers *et al.*, 1986). In explant inoculation the frequency of transformation can be enhanced by addition of acetosyringone to the bacterial culture prior to the inoculation (Sheikholeslam and Weeks 1987). The time taken for transgenic tissue to be produced depends on the culture conditions, the type of explant used and the species of plant from which the explant is taken. Recently, a technique has been developed which substantially reduces the time taken to regenerate transgenic plants that can produce flowers and set seeds (Trinh *et al.*, 1987). In order to do this, epidermal peels from the flowering branches of *Nicotiana plumbaginifolia* are inoculated with *Agrobacterium* and cultured under conditions that induce callus growth and shoot formation which is rapidly followed by production of flower buds. The flowers are fertile and set seed with the whole process taking 8 weeks.

Inoculation of Germinating Seeds
Both of the above methods involve tissue culture techniques which, although in themselves are not relatively complex, can be time consuming and require some specialist laboratory facilities. Moreover, because tissue culture is involved, a certain amount of somaclonal variation may be observed in the regenerated plants which may prove a problem in assessing possible mutants following attempts to use *Agrobacterium*-mediated DNA transfer in *in vivo* mutagenesis (see Chapter 8). A recent report suggests that these difficulties might be overcome by the transformation of germinating seeds (Feldmann and Marks 1987). In this example, seeds of *Arabidopsis thaliana* were imbibed for 12 h and then incubated with a fresh culture of *Agrobacterium* containing a disarmed plasmid vector with a kanamycin resistance gene linked to a Ti plasmid-derived promoter. Plants were selfed and the resulting seed grown on kanamycin to select transformants. The mechanism by which transformation takes place is not known but it appears that the meristematic tissue is not transformed presumably because the *Agrobacterium* cannot obtain access to it. Hence, the transformed tissue was not initially homozygous and there was a large number of insertion events although in future generations the kanamycin resistance was inherited in a Mendelian manner.

Induction of Hairy Roots
Hairy roots can be induced on a variety of plants simply by the inoculation of wounded stem tissue. Generally, seedlings are grown aseptically and a fresh culture

of *A. rhizogenes* is applied to cut stem sections. Within 2–3 weeks, hairy roots proliferate at the site of inoculation and these can be excised and cultured further on media which promote the formation of embryos in the presence of Claforan to remove any remaining bacteria. Whole plants can be regenerated from the embryos under appropriate culture conditions. With co-integrative vectors the plants obtained will contain either the T_R DNA and T_L DNA, or T_R DNA or T_L DNA alone. With binary vectors, the sequence on the vector might be present with or without T_R DNA or T_L DNA or both (Trulson *et al.*, 1986).

Confirmation of Transformation

Confirmation that a plant is transformed can be obtained in several ways, depending on the vector that is adopted, but it is essential that several independent tests are carried out and these can include:

   (i)   a phenotypic assay
  (ii)   an enzymatic assay
 (iii)   Southern blot analysis
  (iv)   analysis of progeny

   Phenotypic assays can mean a plant with an ability to grow under dominant selection (e.g. in the presence of kanamycin or hygromycin). For Ri plasmid vectors where the T_L DNA is present in some plant species, transformed roots display the typical hairy root phenotype: lack of geotropism, and extensive lateral root formation with regenerated plants having wrinkled leaves (Ooms *et al.*, 1985; Guerche *et al.*, 1987). However, it is important to note that these may not be at all apparent in some plant species. Once potential transformants have been selected it is important to assess whether the foreign DNA in the tissue is expressed and this involves an enzyme assay of a genetic marker, (e.g. *nos* or CAT). In order to do this young rapidly expanding tissue is generally used, although it is important to bear in mind that different genes might be expressed to different levels in different tissue. The organization of the foreign DNA in the plant genome can be assessed by Southern blot analysis. Generally it is useful to probe the internal sequences to confirm that all of the DNA has been transferred to the genome and then analyse the border sequences to assess the copy number of the insert and check if it is present as a tandem dimer.

   As shall be seen in Chapter 6, a large amount of variation in the levels of expression of particular gene constructs has been observed in transgenic plants derived by *Agrobacterium*-mediated transformation. In some cases this might be attributed to studying gene expression in tissues at different developmental stages. However, generally, differences in gene expression routinely ascribed to 'position effects' in the plant genome have not been adequately investigated. Because of this it is important that experiments to assess levels of gene expression are carried out on a number of plants so an average result is obtained and that levels of expression are correlated with the levels of expression from a known constitutively expressed marker gene.

References

An, G. (1985). 'High efficiency transformation of cultured tobacco cells', *Plant Physiology* **79**, pp. 568–570.

An, G. (1986). 'Development of plant promoter expression vectors and their use for analysis of differential activity of nopaline synthase promoter in transformed tobacco cells', *Plant Physiology* **81**, pp. 86–91.

An, G., Watson, B.D., Stachel, S., Gordon, M.P. and Nester, E.W. (1985). 'New cloning vehicles for the transformation of higher plants', *EMBO J.* **4**, pp. 277–284.

Bevan, M.W. (1984). 'Binary *Agrobacterium* vectors for plant transformation', *Nucleic Acids Res.* **12**, pp. 8711–8721.

Colau, D., Negrutiu, I., Van Montagu, M. and Hernalsteens, J.P. (1987). 'Complementation of a threonine dehydratase-deficient *Nicotiana plumbaginifolia* mutant after *Agrobacterium tumefaciens*-mediated transfer of *Saccharomyces cerevisiae* ILV1 gene', *Mol. Cell. Biol.* **7**, pp. 2552–2557.

Crawford, N.M., Campbell, W.H. and Davis, R.W. (1986). 'Nitrate reductase from squash, cDNA cloning and nitrate regulation', *Proc. Nat. Acad. Sci. USA* **83**, pp. 8073–8076.

DeBlock, M., Herrera-Estrella, L., Van Montagu, M., Schell, J. and Zambryski, P. (1984). 'Expression of foreign genes in regenerated plants and their progeny', *EMBO J.* **3**, pp. 1681–1689.

DeBlock, M., Batterman, J., Vandewiele, M., Dockx, M., Theon, C., Gossele, V., Rao Movva, N., Thompson, C., Van Montagu, M. and Leemans, J. (1987). 'Engineering herbicide resistance into plants by expression of a detoxifying enzyme', *EMBO J.* **6**, pp. 2513–2518.

Fillatti, J.J., Sellmer, J., McCown, B., Haissig, B. and Comai, L. (1987). '*Agrobacterium*-mediated transformation of *Populus*', *Mol. Gen. Genet.* **206**, pp. 192–199.

Feldmann, K.A. and Marks, M.D. (1987). '*Agrobacterium*-mediated transformation of germinating seeds of *Arabidopsis thaliana*: A non-tissue culture approach', *Mol. Gen. Genet.* **208**, pp. 1–9.

Fraley, R.T., Rogers, S.G., Horsch, R.B., Eichholtz, D.A., Flick, C.L., Hoffman, N.L. and Saunders, P.R. (1985). The SEV system: A new disarmed Ti plasmid vector system for plant transformation. *Bio/Technology* **3**, pp. 629–635.

Guerche, P., Jouanin, L., Tepfer, D. and Pelletier, G. (1987). 'Genetic transformation of oilseed rape (*Brassica napus*) by the Ri plasmid of *Agrobacterium rhizogenes* and analysis of the transformed phenotype', *Mol. Gen. Genet.* **206**, pp. 382–386.

Helmer, G., Casadaban, M., Bevan, M., Kayes, L. and Chilton, M-D. (1984). 'A new chimeric gene as a marker for plant transformation: The expression of *Escherichia coli* β-galactosidase in plant cells', *Bio/Technology* **2**, pp. 520–527.

Herrera-Estrella, L., DeBlock, M., Messens, E., Hernalsteens, J-P., Van Montagu, M. and Schell, J. (1983a). 'Chimeric genes as dominant selectable markers in plant cells', *EMBO J.* **2**, 987–995.

Herrera-Estrella, L., Depicker, A., Van Montagu, M. and Schell, J. (1983b). 'Expression of chimeric genes transferred into plant cells using a Ti plasmid-derived vector', *Nature* **308**, pp. 209–213.

Hille, J., Roelvink, A., Franssen, H., Van Kammen, A. and Zabel, P. (1986). 'Bleomycin resistance: a new dominant selectable marker for plant cell transformation', *Plant Mol. Biol.* **7**, pp. 171–176.

Hoekema, A., Hirsch, P.R., Hooykaas, P.J. and Schilperoot, R.A. (1983). 'A binary plant vector strategy based on the separation of the *vir* and T-region of agrobacteria', *Nature* **303**, pp. 179–181.

Inze, D., Follin, A., Velten, J., Velten, L., Prinsen, E., Rudelsheim, P., Van Onckelen, H., Schell, J. and Van Montagu, M. (1987). 'The *Pseudomonas savastonoi* tryptophan-2-monooxygenase is biologically active in *Nicotiana tabacum*', *Planta* **172**, pp. 555–562.

Jefferson, R.A., Kavanagh, T.A. and Bevan, M.W. (1987). 'GUS fusions: β-glucuronidase as a sensitive and versatile gene fusion marker in higher plants', *EMBO J.* **6**, pp. 3901–3907.

Jones, J.D.G., Svab, Z., Harper, E.C., Horwitz, C.D. and Maliga, P. (1987). 'A dominant nuclear streptomycin resistance marker for plant cell transformation', *Mol. Gen. Genet.* **210**, pp. 86–91.

Komari, T., Halperin, W. and Nester, E.W. (1986). 'Physical and functional map of supervirulent *Agrobacterium tumefaciens* tumor-inducing plasmid pTiBo542', *J. Bact.* **166**, pp. 88–94.

Koncz, C., Olsson, O., Langridge, W.H.R., Schell, J. and Szalay, A.A. (1987). Expression and assembly of functional bacterial luciferase in plants', *Proc. Nat. Acad. Sci. USA* **84**, pp. 131–135.

Lefebvre, D.D., Miki, B.L. and Laliberte, J-F. (1987). 'Mammalian metallothionein functions in plants', *Bio/Technology* **5**, pp. 1053–1956.

Marton, L., Wullems, G.J., Molendijk, L. and Schilperoot, R.A. (1979). '*In vitro* transformation of *Nicotiana tabacum* by *Agrobacterium tumefaciens*', *Nature* **277**, pp. 129–131.

Meyer, P., Heidmann, I., Forkmann, G. and Saedler, H. (1987). 'A new petunia flower colour generated by transformation of a mutant with a maize gene', *Nature* **330**, pp. 677–678.

Müller, A.J. (1983). 'Genetic analysis of nitrate reductase-deficient tobacco plants regenerated from mutant cells. Evidence for duplicate structural genes', *Mol. Gen. Genet.* **192**, pp. 275–281.

Ooms, G., Karp, A., Burrell, M.M., Twell, D. and Roberts, J. (1985). 'Genetic modification of potato development using Ri T-DNA', *Theor. Appl. Gen.* **70**, pp. 440–446.

Otten, L.A.B.M. and Schilperoot, R.A. (1978). 'A rapid microscale method for the detection of lysopine and nopaline dehydrogenase activities', *Biochim. Biophys. Acta* **527**, pp. 497–500.

Ow, D., Wood, K.V., DeLuca, L., DeWet, J.R., Helsinki, D. and Howell, S.H. (1986). 'Transient and stable expression of firefly luciferase gene in plant cells and transgenic plants,' *Science* **234**, pp. 856–859.

Pietrzak, M., Shillito, R., Hohn T. and Potrykus, I. (1986). 'Expression in plants of two bacterial antibiotic resistance genes after protoplast transformation with a new plant expression vector', *Nucleic Acids Res.* **14**, pp. 5857–5868.

Pua, E-C., Meyra-Palta, A., Nagy, F. and Chua, N-H. (1987). 'Transgenic plants of *Brassica napus* L.', *Bio/Technology* **5**, pp. 815–817.

Reiss, B., Sprengel, R., Will, H. and Schaller, H. (1984). 'A new sensitive method for qualitative and quantitative assay of neomycin phosphotransferase in crude cell extracts', *Gene* **30**, pp. 211–218.

Rogers, S.G., Horsch, R.B. and Fraley, R.T. (1986). 'Gene transfer to plants: production of transformed plants using Ti plasmid vectors', in *Methods in Enzymology* **118**. Eds. Weissbach, A. and Weissbach, H., pp. 627–640. London, Academic Press.

Schell, J. (1987). 'Transgenic plants as tools to study the molecular organisation of plant genes', *Science* **237**, pp. 1176–1183.

Scott, R.J. and Draper, J. (1987). 'Transformation of carrot tissues derived from preembryogenic suspension cells: A useful model system for gene expression studies in plants', *Plant Mol. Biol.* **8**, pp. 265–274.

Shah, D., Horsch, R.B., Klee, H.J., Kisherer, G.M., Winter, J.A., Turner, N.E., Hironaka, C.M., Sanders, P.R., Gasser, C.S., Aykent, S., Siegel, N.R., Rogers, S.G. and Fraley, R.T. (1986). 'Engineering herbicide tolerance in transgenic plants', *Science* **233**, pp. 478–481.

Sheikholeslam, S.N. and Weeks, D. (1987). 'Acetosyringone promotes high efficiency transformation of *Arabidopsis thaliana* explants by *Agrobacterium tumefaciens*', *Plant Mol. Biol.* **8**, pp. 291–298.

Simpson, R.B., Spielmann, A., Margossian, L. and McKnight, T.D. (1986). 'A disarmed binary vector from *Agrobacterium tumefaciens* functions in *Agrobacterium rhizogenes*', *Plant Mol. Biol.* **6**, pp. 403–415.

Stougaard, J., Abildsten, D. and Marcker, K.A. (1987). 'The *Agrobacterium rhizogenes* pRi TL-DNA segment as a gene vector system for the transformation of plants', *Mol. Gen. Genet.* **207**, pp. 251–255.

Stougaard Jensen, J., Marcker, K.A., Otten, L. and Schell, J. (1986). 'Nodule-specific expression of a chimeric soybean leghemoglobin gene in transgenic *Lotus corniculatus*', *Nature* **321**, pp. 669–674.

Trinh, T.H., Mante, S., Pua, E-C. and Chua, N-H (1987). 'Rapid production of transgenic flowering shoots and F1 progeny from *Nicotiana plumbaginifolia* epidermal peels', *Bio/Technology* **5**, pp. 1081–1084.

Trulson, A.J., Simpson, R.B. and Shahin, E.A. (1986). 'Transformation of cucumber (*Cucumis sativus* L.) plants with *Agrobacterium rhizogenes*', *Theor. Appl. Genet.* **73**, pp. 11–15.

Van Haute, E., Joos, H., Maes, S., Warren, G., Van Montagu, M. and Schell, J. (1983). 'Intergeric transfer and exchange recombination of restriction fragments cloned in pBR322: a novel strategy for reversed genetics of the Ti Plasmids of *Agrobacterium tumefaciens*', *EMBO J.* **2**, pp. 411–418.

Waldron, C., Murphy, E.B., Roberts, J.L., Gustafson, G.D., Armour, S.L. and Malcolm, S.K. (1985). 'Resistance to hygromycin B', *Plant Mol. Biol.* **5**, pp. 103–108.

Zambryski, P., Joos, H., Genetello, C., Leemans, J., Van Montagu, M. and Schell, J. (1983). 'Ti plasmid vector for the introduction of DNA into plant cells without alteration of their normal regeneration capacity', *EMBO J.* **2**, pp. 2143–2150.

Chapter 4

Vectors Based on Viral Genomes

The Potential of Vectors Based on Viral Genomes

Whilst vectors have been developed from the Ti plasmid which utilize the ability of *Agrobacterium* to insert DNA into the genome of plant cells, this system, despite many recent improvements, still has some inherent limitations. These include the relatively complex experimental manipulations that are involved in engineering, both *in vitro* and *in vivo*, the foreign DNA to be inserted into the plant genome, as well as the difficulty that, once within the plant cell, the copy number of the transferred DNA remains low, requiring refined detection techniques in order to study the expression of the foreign DNA sequences. Moreover, while difficulties in plant regeneration in many plant species, not least the cereals, remain, the transformation of these plants using *Agrobacterium*-based vectors is impractical. In order to bypass some of these difficulties a great amount of attention has been paid to plant viruses as potential vectors. This interest is derived from the fact that viruses as pathogenic agents normally enter the plant cell, express the information contained in their genomes and can replicate to achieve a high copy number. Thus, by their very nature, plant viruses have several characteristics not shared by the Ti plasmid which could prove helpful in their use as vectors. Broadly, these characteristics can be summarized as follows.

Host Range
The host ranges of plant viruses can include the economically important crops such as the cereals which, because of difficulties encountered with tissue culture, may not be amenable to transformation using vectors based on the Ti plasmid.

Ease of Insertion into the Plant Cell
Although most plant viruses are transmitted by insects in nature, many can be mechanically inoculated onto wounded plant tissue and result in systemic infection. Moreover, in a few cases either virion DNA (or RNA), viral DNA which has been cloned (or a cDNA representing viral RNA), or RNA derived from cloned DNA can be infectious.

Constitutive Expression
The promoters of viral genes are generally considered to be constitutive. Hence viral genes may be expressed in many different cell types from tissue at different developmental stages. Some promoters may be considered as 'strong', giving rise to high levels of specific viral transcripts; however, caution is needed in the designation of a particular viral promoter as being 'strong' because high levels of an individual transcript can also result from a high copy number of the gene encoding it as well as the stability of the transcript.

Replication and Systemic Passage
Viruses replicate within plant cells to achieve high copy number and progeny virions can pass throughout the host plant so that although viral DNA is not generally integrated into the plant genome, expression of the viral DNA is likely to take place in different tissues of the infected plant. Moreover, in some rare cases of viral infection, the viral nucleic acid can passage throughout the infected plant without being packaged within a virion particle.

Ease of Manipulation
The genomes of many plant viruses are relatively small and once cloned in bacterial vectors may be manipulated easily in *E. coli* prior to re-introduction into the plant by inoculation.

These features however are not shared by all plant viruses and the inherent characteristics of a particular virus, for example its genome structure and mode of expression, can impose restrictions on its use as a vector and hence the development of vectors based on viral genomes has been relatively limited. In order to successfully develop this type of system several basic criteria need to be met. Firstly, the virus must be easily inoculated onto test plants and the resulting infection produce characteristic symptoms. This eases the analysis of both the viral infection process and the expression of viral nucleic acid. Obviously the shorter the time taken for symptoms to appear the better. Secondly, the nucleic acid of the viral genome, once isolated and cloned in a bacterial cloning vector, must be able to be re-introduced into the plant and remain infectious. This allows not only characterization of the infectious nucleic acid by sequencing but also the manipulation of the genome *in vitro*. Thus *in vitro* mutagenesis can be carried out to assess which portions of the viral genome can be modified and how much foreign DNA can be stably inserted. Thirdly, it is important that the nucleic acid of the viral genome can be manipulated to some extent, allowing the insertion of foreign DNA, without interfering with its ability to replicate and express itself. This is particularly important if systemic passage of the modified viral genome throughout the whole plant is required. Fourthly, the viral genome needs to be a stable replicon and able to accumulate in the plant cell. The replicative functions may be encoded by the virus or the host but either might be dependent on a particular stage of the cell cycle of the infected plant. If this is the case, replication might be restricted to dividing cells (i.e. meristems) whereas, if viral replication is not associated with cell division, it might be expected that large amounts of viral DNA would accumulate in all plant cells. In the following descriptions of individual plant viruses and how some of them have been adapted to transfer foreign DNA into plant cells, not only the advantages but also some of the limitations inherent in their use as vectors will be illustrated.

RNA Viruses as Vectors

The Structure of RNA Viruses

The majority of plant viruses that have been characterized comprise of a single-stranded RNA genome. The different RNA viruses can vary according to their particle structure (spherical or rod-like) and whether they are segmented or non-segmented. The genomes of most RNA viruses are of the same polarity as mRNA and can be translated directly on the ribosomes of the infected cell to produce viral specific polypeptides. Amongst RNA viruses there is a variety of strategies to translate their RNA genomes into proteins, as follows:

(i) The viral genomic RNA serves as a monocistronic mRNA

(ii) The viral genomic RNA is a monocistronic mRNA for a polyprotein

(iii) Expression of viral genomic RNA is limited to the 5' proximal gene and involves synthesis of subgenomic mRNAs to express their viral products

(iv) Expression of the viral genomic RNA involves synthesis of a polyprotein followed by proteolytic processing as well as the synthesis and translation of subgenomic RNAs (Dougherty and Hiebert 1986).

Obviously the mechanism of gene expression existing in a particular virus needs to be considered when contemplating its potential use as a vector. For example, it may prove more difficult to insert foreign sequences into RNA and obtain correct expression if the RNA encodes a polyprotein or acts as a polycistronic mRNA. It may be easier to insert a foreign sequence into a subgenomic RNA, provided the subgenomic RNA or its gene product can be modified without interfering with viral gene expression or function. So that the techniques of recombinant DNA technology can be applied to the genomes of the RNA viruses, the first step is to convert the RNA into a DNA. This is carried out by synthesizing a complementary copy of DNA (cDNA) to the viral RNA followed by synthesis of the second strand of DNA. The double-stranded DNA can then be cloned in *E. coli*. It might be expected that the resulting DNAs may themselves be infectious, particularly if normal viral replication proceeds through a DNA intermediate or if an infectious RNA can be synthesized from the cDNA within the plant cell. This has been found to be the case with the bacteriophage Q_β, poliovirus and with potato spindle tuber viroid (PSTV). Viroids are extremely small, unencapsidated, single-stranded covalently-closed RNA molecules and clones representing mono-meric and dimeric genomes are infectious. In this case infection is thought to result from transcription *in vivo* (Cress *et al.*, 1983). An alternative is to use the cloned DNAs to synthesize RNA *in vitro* and inoculate the plant tissue with this.

Tobacco Mosaic Virus (TMV) as a Vector
The approach of using a cDNA representing a viral genome to direct the synthesis of infectious RNA *in vitro* has been used with Tobacco Mosaic Virus (TMV) which has been engineered to contain the CAT gene (Takamatsu *et al.*, 1987).

TMV is a single-component, messenger-sense, single-stranded RNA virus. It encodes four polypeptides, the 130K and 160K polypeptides which are translated directly from the genomic RNA using the same initiation codon and the 30K and coat protein which are translated from subgenomic RNAs (Fig. 4.1). It is thought that the 130K and 160K polypeptide are involved in the replication of the virus whereas the 30K polypeptide is involved in cell-to-cell transfer. Mutants defective in the coat protein can be propagated in the plant suggesting that the coat protein is not involved directly in viral RNA replication and that this region might be used as a site for the insertion of foreign DNA. Moreover, the coat protein accumulates to very high levels in the infected plant and hence high levels of the foreign gene product might also accumulate. In order to assess whether infectious chimeric cDNAs could be constructed, a cDNA clone representing TMV was prepared from which the coat protein gene was removed (pLDC529) (Fig. 4.1) and this was inserted into a bacterial cloning vector downstream from a bacterial promoter sequence. RNA representing the cDNA was synthesized and re-constituted into virion particles *in vitro* in the presence of coat protein and used to inoculate test plants. Although the RNA was unable to systematically passage throughout tobacco it could replicate in the inoculated leaf and the subgenomic RNA which would under normal circumstances encode the coat protein also accumulated. Chimeric genomes containing the CAT gene inserted in place of the coat protein (pCL29) (Fig. 4.1) were also able to replicate, producing a subgenomic mRNA with large levels of CAT activity being detected in the inoculated tissue. This work demonstrated that a foreign gene could be introduced into whole plants using an RNA virus vector and although it was unable to passage throughout the plant, expression of the foreign gene could be obtained at high levels in the inoculated tissue.

Brome Mosaic Virus (BMV) as a Vector
Obviously for multi-component viruses each of the cloned components may need to be introduced into the same cell and expressed correctly for infection to proceed. This approach has been used to engineer Brome Mosaic Virus (BMV) to express CAT in infected barley protoplasts (French *et al.* 1986). BMV is a tripartite RNA virus and its principal hosts are monocotyledonous plants. The three genomic RNAs are packaged separately and contain a 5′ cap and a 3′ tRNA-like structure which are thought to be important in their expression and replication. RNAs 1 and 2 encode single polypeptides whereas RNA 3 encodes a 35 000 molecular weight polypeptide and a subgenomic RNA, RNA 4, which is

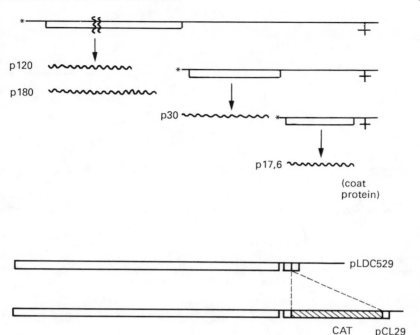

Fig. 4.1 The structure of the TMV genome. The genomic and subgenomic RNAs are shown as well as their protein products. The cDNA clone representing the genome with the deleted coat protein gene is shown, pLDC529, as is the site of the insertion of the CAT gene in pCL29.

probably derived by partial transcription of RNA 3. RNA 4 encodes the viral coat protein which can accumulate to high levels in infected tissue (Fig. 4.2). cDNA copies of each of the three RNA components of the virus have been made and were cloned into bacterial plasmids just downstream from a modified bacterial promoter (pB1, pB2, and pB3 representing RNAs 1, 2, and 3, respectively) allowing the production *in vitro* of RNA transcripts representing each of the viral RNAs. This system was used to test the ability of single RNAs as well as combinations of RNAs, to initiate replication following inoculation of barley protoplasts. Only RNAs 1 and 2 are required for viral RNA replication in protoplasts whereas all three RNAs are required to initiate infection in plants. This suggests that the RNA 3 gene products are involved in the systemic transfer of the virus in plants and that they are not necessary for the replication of RNAs 1 and 2. In order to investigate whether the viral coat protein was required for replication of its own RNA the coat protein gene in pB3 was deleted and the resultant truncated RNA was used in inoculations with RNAs 1 and 2. It was found that the modified RNA was replicated in protoplasts, albeit at a three-fold reduced level. This indicated that the coat protein gene could be replaced by sequences of foreign nucleic acid without abolishing its own replication. The CAT

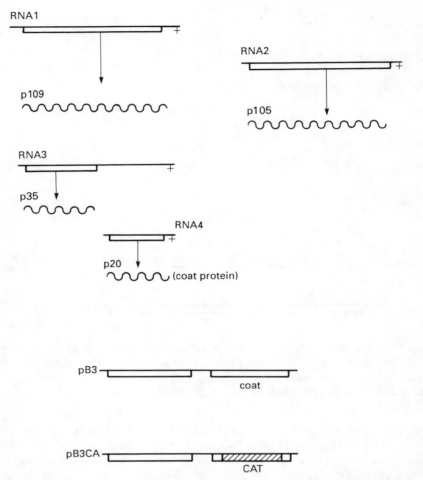

Fig. 4.2 The structure of the BMV genome. The four RNAs of the BMV each encode a polypeptide the smallest of which, p20, is the coat protein. A cDNA, representing RNA 3, was cloned forming the plasmid pB3, this in turn was engineered to contain a CAT gene within the region encoding the coat protein pB3CA.

gene was inserted in place of the coat protein gene in pB3 and the RNA synthesized *in vitro* and used to inoculate protoplasts along with RNAs 1 and 2. The chimeric RNAs as well as the modified RNA 4 accumulated in the infected protoplasts although once again at reduced levels when compared with accumulation of the wild type RNAs. In addition, the transfected cells contained high amounts of CAT enzyme activity. Indeed, the levels of enzyme activity observed were particularly high when compared with levels seen when analogous constructs were inserted into plant cells using Ti plasmid-derived vectors. This is probably due to both the high copy number of the gene sequences in the cells inoculated with the viral RNA constructs and the different promoters used to direct synthesis from the CAT gene. It is also of interest that in these experiments constructs were made which contained the CAT gene in different positions with regard to the promoter sequence of the coat protein gene. In one instance a construct contained a translation termination codon positioned between the first AUG of the subgenomic RNA and the CAT gene initiation codon. In this case the level of CAT activity was reduced by only 35% when compared with other constructs in which the intervening termination codon was not present. Similar results were obtained with the CAT construct in TMV. This demonstrates that internal initiation and re-initiation of translation can take place at significant levels in plant cells and provides an example of how this system can be used to investigate some of the fundamental questions concerning the control of gene expression.

A Potential Limitation of Using RNA Viruses as Vectors
The mechanism by which RNA viruses replicate and synthesize their subgenomic RNAs has not been totally resolved. It is likely that synthesis of a second strand of RNA is carried out by RNA-dependent RNA polymerases, which could be encoded either by the host or the virus. It has been argued that engineered sequences within RNA viral genomes may not be stable due to errors which are known to take place during RNA-dependent RNA synthesis. This could cause difficulties in using engineered viral genomes to systematically infect plants a process which requires multiple rounds of replication to take place. With BMV the RNA transcripts transcribed *in vitro* are also infectious when inoculated onto whole plants so this system could be used to assess the stability of the foreign DNA sequences under such conditions.

DNA Viruses as Vectors

There are only two groups of plant viruses currently known that contain a genome comprising of DNA. These are the caulimovirus family and the gemini viruses. Obviously, with their DNA genomes being immediately amenable to recombinant DNA technology, these viruses have been extensively studied, not least because of their potential as vectors.

Caulimoviruses as Potential Vectors

The Structural Organization of Caulimoviruses
The caulimoviruses fulfil several of the critera required if a virus is to be a potential vector. In particular both the virus and the virion DNA can be inoculated onto wounded test plants to produce characteristic symptoms within a few weeks and this has eased their study greatly.

The caulimovirus group of viruses contain a double-stranded circular DNA genome of about 8000 bp and the group comprises of several members, all of which are similar in their mode of transmission in the type of inclusion body which accumulates in infected tissue and in their virion morphology. In addition they are serologically related. Cauliflower Mosaic Virus (CaMV) is the type virus of the group, infects members of the genus *Brassica* and has been the most extensively studied caulimovirus. The viral DNA is packaged in an icosahedral virion particle of about 50 nm diameter and molecular weight 22.8×10^6 daltons. Within the virion particle a large proportion of the DNA is twisted and knotted and there are site-specific sequence discontinuities in each strand of the DNA. The number of these discontinuities varies between isolates of CaMV as well as between different members of the caulimovirus group. Each viral isolate has one strand of DNA, the α-strand, containing a single discontinuity whilst two interruptions in the complementary strand subdivide it into the β-, γ- and, if a third interruption is present, the δ-strand. Originally it was thought that the discontinuities were gaps but sequencing of virion DNA demonstrated that they were, in effect, flaps comprising of regions of sequence overlap of either 8, 15 or 19 nucleotides, depending on the position of the flap. The discontinuities probably arise during replication of the viral DNA. However, neither the twisted conformation nor the discontinuities are necessary for initiation of infection since cloned linear double-stranded viral DNA, when inoculated on to test plants, is infectious.

The Sequence of the Caulimovirus Genome

The isolates of CaMV that have been sequenced share the same general organization of potential protein coding sequences. There are six large (I–VI) and two small (VII and VIII) open reading frames (Fig. 4.3). The nucleotide sequences vary between different isolates by 5% but the sequences of the two largest intergenic regions between ORFs V and VI and ORFs VI and VII are highly conserved suggesting that they may play important roles in the initiation and termination of transcription. The DNA sequence of Carnation Etched Ring Virus (Hull *et al.*, 1986) and Figwort Mosaic Virus (Richins *et al.*, 1987), two other members of the caulimovirus group, have been obtained and their overall genome organization is similar to CaMV, although no similarities with ORFs VII and VIII were found. Only one strand of CaMV DNA, the α-strand, is transcribed. The ORFs are arranged on the α-strand in alternate reading frames and generally are either close to each other or overlap. Most of the potential polypeptides encoded by the virus have been identified in infected tissue although their function during viral replication cycle remains largely unknown. Nevertheless, insertion mutagenesis indicates that ORFs I, III, IV, V and VI are required for the development of normal infection as judged by the development of viral symptoms on test plants. For some time it has been known that naturally-occurring variants containing a deletion in ORFII cannot be transmitted by aphids although the infection of host plants remains unimpaired. Hence the 18 000 molecular-weight polypeptide encoded by ORFII has been termed the 'aphid transmissibility factor'. The polypeptide is associated with the inclusion bodies which develop in the infected cell but how its presence permits aphid transmission is still unknown. Comparison of the DNA sequence with the amino

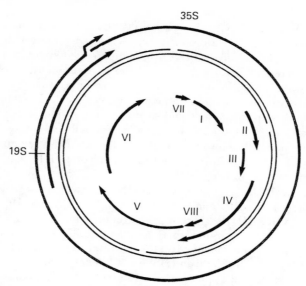

Fig. 4.3 The organization of the CaMV genome. The circular DNA of the CaMV genome is shown with the sites of the three discontinuities. The ORFs are depicted as arrows displaying their polarity and numbered I–VIII. The two most abundant transcripts the 19S and 35S RNAs are shown.

acid content of the coat protein suggests that ORFIV encodes a coat protein precursor and this proposal has support from the finding that a polypeptide synthesized in *E. coli* from a plasmid containing ORFIV reacts with antisera raised against viral coat protein. ORFV is the most highly conserved open reading frame between isolates of CaMV with the capacity to encode a 79 000 molecular weight polypeptide. There is considerable homology between the sequence of ORFV and the *pol* genes (i.e. the genes coding for RNA-dependent DNA polymerase or reverse transcriptase) of animal retroviruses; furthermore, yeast cells containing ORFV cloned in an expression vector accumulate reverse transcriptase activity. This suggests that ORFV encodes a reverse transcriptase which probably is involved in the replication of the viral genome. Of the viral proteins that accumulate in infected plants, the product of ORFVI, the 66 000 molecular-weight inclusion body protein, is by far the most abundant. The function of the inclusion body protein is not known although it does play a role in normal symptom expression as well as being involved in the systemic passage of the virus throughout the infected plant. The inclusion bodies themselves are thought to be the sites of virion assembly. As yet it has not been demonstrated directly that ORFs VII and VIII encode polypeptides in infected tissue and ORFVII can be deleted without destroying viral replication and symptom expression (Covey and Hull 1985).

Transcription of the CaMV Genome

Transcription of CaMV DNA takes place in the nucleus of the plant cell and its sensitivity to α-amanitin indicates the host cell RNA polymerase-II is responsible. Within the nucleus of the infected cell closed, supercoiled circular viral genomes which are considered to be transcription complexes, accumulate. These molecules associate with histone to form minichromosomes and interestingly, do not contain the site-specific discontinuities thus removing the need for the RNA polymerase to traverse the single gap in the α-strand during transcription. The minichromosomes are reminiscent of the SV40 minichromosomes which accumulate in infected mammalian cells. However, unlike SV40 there is no evidence to suggest that CaMV DNA integrates into the nuclear genome of the infected plant and the virus is not transmitted to progeny.

A variety of CaMV transcripts have been detected in infected tissue. The most abundant of these are the 19S and 35S transcripts (Fig. 4.3). The 19S transcript is the mRNA of ORFVI, the inclusion body protein. The DNA sequences surrounding ORFVI are similar to those surrounding eucaryotic genes and presumably contain regions that control transcription. The 19S RNA is 5′-capped and has a poly(A) tail. The 35S transcript is a full-length transcript of the CaMV genome with an overlap of approximately 180 bases at its 3′ and 5′ ends. Hence during transcription, RNA polymerase traverses the transcriptional stop site at the position of the 3′ terminal on its first circuit of the genome whereas it terminates at the same signal during the second passage. Either leaky transcription termination or the secondary structure of the DNA or nascent RNA may be responsible for this. As with the 19S RNA the upstream sequences from the transcriptional start site of the 35S RNA resemble those of other eucaryotic promoter sequences and these have been studied in detail using Ti plasmid vectors (Chapter 6). The function of the 35S RNA has aroused much interest. There have been suggestions that it may act as a polycistronic mRNA and it is now generally considered that it serves as a template for reverse transcription to produce DNA during the replication of the virus. The suggestion that the 35S RNA may serve as a polycistronic mRNA has arisen because although discrete mRNAs for ORFs V and VI have been detected in infected tissue this has not been the case for ORFs I–IV. Because ORFs I–IV are either very close to each other or overlap, a model has been proposed in which the ribosomes, during translation, move along the RNA from one ORF to the next without disengaging from the RNA. However, although *in vitro* mutagenesis indicates that the translation of ORFs VII and I may be linked, the evidence for linked translation of ORFs I–IV remains indirect, being based on the structure of viral genomes arising from the production of viable deletion mutants *in vivo* following the inoculation of test plants with modified viral genomes.

Replication of CaMV

It is now generally held that CaMV replicates via reverse transcription of the 35S RNA (Hull and Covey 1983) (Fig. 4.4). The model to describe the replication process has been derived from observations of the types of putative replication intermediates which accumulate in infected cells, the structure of the 35S RNA,

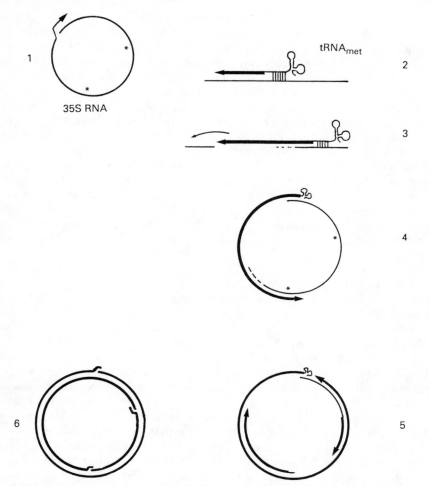

Fig. 4.4 The replication of the CaMV genome. The 35S RNA (1) serves as template for reverse transcription which is initated by a tRNA$_{met}$ hybridizing to the 35S RNA (2). DNA synthesis, represented by the thick line (3) proceeds, traverses the gap in the RNA template which in turn is degraded by an RNase H-like activity. DNA synthesis proceeds around the genome (4) and second strand synthesis is primed by RNase H-resistant regions of RNA and procedes to form a gapped genome (6).

the processing of viral RNA engineered to contain intron sequences, the observation that ORFV probably encodes a reserve transcriptase and the detection of this activity in infected plant cells (Thomas *et al.*, 1985). The model proposes that tRNA$_{met}$, which contains a region of complementarity to the 35S RNA, acts as a primer for the synthesis of DNA from the 5′ region of the 35S RNA by the reverse transcriptase activity encoded by ORFV. The terminal repeat sequence in the 35S RNA allows the newly synthesized DNA to recognize the complementary sequence at the 3' region of the 35S RNA and hence the reverse transcriptase traverses the gap and continues synthesis around the RNA. An RNAse H activity is presumed to digest the RNA template following DNA synthesis except at sites of polypurine-rich sequences located near the strand-specific gaps in the genome and these serve to prime second strand synthesis of DNA. Synthesis of the second strand proceeds until the next primer is reached and the reverse transcriptase displaces the primer to produce the site-specific discontinuities (Covey and Hull 1985).

This mode of replication is likely to cause some difficulties in producing a vector based on the CaMV genome because in order to be a stable replicon several specific sites on the viral genome are required as well as a virus-specific enzyme activity. In addition, replication based on an RNA intermediate may be more prone to errors than replication based on DNA. Moreover, any sequence either containing an intron, or sequences similar to intron/exon borders, are likely to be spliced.

The Infectivity of the Cloned CaMV Genome

Cloned CaMV DNA is infectious when inoculated onto wounded test plants. If the viral DNA is cloned as a monomer, the bacterial cloning vector must be excised prior to inoculation whereas if cloned as a dimer, or partial dimer, excision of the plasmid is not necessary. In the latter case infection can result from either intramolecular recombination or the synthesis of the 35S RNA which can serve as the replication intermediate. As we have already seen, only ORFs II and VII can be modified without destroying normal viral infectivity. This suggests that foreign DNAs could be inserted onto those regions without disrupting viral functions. However, sequences of DNA greater than 250 bp inserted into the viral genome are not stable (Gronenborn *et al.*, 1981). This may be a reflection of both the stability of the viral genome and packaging constraints imposed by the virion particle. In order to overcome these constraints, attempts have been made to establish helper virus systems. The rationale behind this is that pairwise inoculation of test plants with mutant viral genomes which individually could not initiate infection might result in the mutants complementing each other so as to produce those conditions required for viral function. Unfortunately this approach has not proved feasible because of extensive recombination taking place between the inoculated mutant DNAs within the plant cell (Walden and Howell 1982).

CaMV as a Vector

Despite the difficulties described above, CaMV has been used to insert foreign genes into whole plants and obtain their correct expression (Brisson *et al.*, 1984; Lefebvre *et al.*, 1987). In order to carry this out, three considerations needed to be borne in mind. Firstly, the foreign DNA had to be inserted at a site in the genome which could be modified. Secondly, the gene needed to be inserted in such a way that it might be expressed and not interfere with the translation strategy of the virus and, finally, the inserted DNA had to be small so that it could be stably

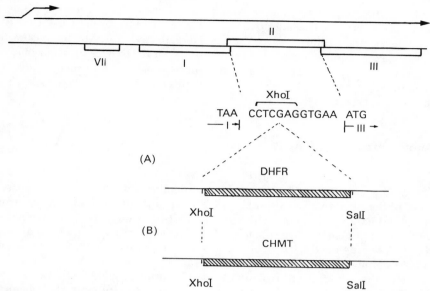

Fig. 4.5 Engineering the CaMV to contain chimeric DNA. The majority of ORFII is deleted. The XhoI site allows the insertion of (A) DHFR or (B) CHMT on an XhoI-SalI fragment.

maintained within the viral genome and packaged into the virion. In these examples, ORFII of CaMV was deleted so that only the first five codons and the stop codon of the gene remained. A unique Xho I site was present immediately in front of the stop codon and into this site the genes encoding either dihydrofolate reductase (DHFR) or Chinese Hamster metallothionein (CHMT) were inserted (Fig. 4.5). The cloned viral genome containing DHFR was inoculated onto wounded test plants and it was found that the DNA containing the insertion was systematically passaged throughout the plant and produced the normal symptoms of CaMV infection. The infected plants accumulated a protein which cross-reacted with antisera raised against the DHFR gene product and DHFR enzyme activity was present in plant extracts. Plants infected with the virus containing the DHFR gene remained relatively unaffected when sprayed with a solution of methotrexate whereas those without the virus containing the gene showed evidence of severe senescence. In the case of CHMT, the inoculated plants contained the metallothionein protein and had an increased cadmium binding property.

Like TMV and BMV, the examples of CaMV engineered as a vector expressed the foreign DNA at very high levels in the inoculated plants. However, in CaMV the foreign DNA inserted was relatively small and systemic infection can only take place in the normal hosts of CaMV, members of the genus *Brassica*. In time these difficulties may be overcome by deleting more regions of the genome, such as ORFVII, so that more sequences are tolerated and by modifying ORFVI so that the host range of the virus might be increased. Hence CaMV, used as a single-component vector in whole plants, has not proven to be very versatile. Nevertheless, these difficulties may not preclude the use of CaMV as a vector for transient expression in protoplasts where only viral DNA replication may be required. Moreover, helper systems may still be constructed in such a way that recombination is reduced or eliminated and this could increase the capability of CaMV as a vector.

Gemini Viruses as Potential Vectors

The Structure, Host Range and Infectivity of Gemini Viruses
The gemini viruses are one of the most economically important plant viruses, being responsible for diseases in a variety of crop plants such as maize, beans, wheat and potato. They comprise a group of viruses which are characterized by the twinned or geminate morphology of the virion particle. Each virion appears to consist of a twinned particle of two incomplete icosahedra of 18–20 nm width and 30–36nm length. They are also characterized by possession of genomes comprising of either one or two circular single-stranded DNAs of between 2500 and 3000 bases. Analysis of both the buoyant density and the chemical composition of one gemini virus with a bipartite genome, Bean Golden Mosaic Virus (BGMV), suggests that a single DNA molecule is individually encapsidated within each geminate particle. The gemini viruses that have been characterized to date can be grouped according to their host range and/or insect vector, that is to say those that are each transmitted by a different species of leaf hopper or those that can be transmitted by a single species of whitefly. The group of viruses that are transmitted by leaf hoppers comprise a single component genome and contain species which can infect monocotyledonous plants (e.g. maize streak virus, MSV, and wheat dwarf virus, WDV) and species which can infect dicotyledonous plants (e.g. beet curly top virus, BCTV). Those viruses transmitted by whitefly comprise a two-component genome and infect dicotyledonous plants (e.g. BGMV, cassava latent virus, CLV, tomato golden mosaic virus, TGMV). The gemini viruses transmitted by whiteflies can generally be mechanically inoculated onto test plants with subsequent infection producing characteristic symptoms. For example, CLV and TGMV can be mechanically inoculated onto species of *Nicotiana* and BGMV onto beans. Hence these viruses have been studied in detail. On the other hand, with the exception of BCTV, those viruses transmitted by leafhoppers cannot be mechanically transmitted. Here infection can only be obtained following inoculation by leaf hoppers which have recently fed on infected plants. The inability to initiate infection by mechanical inoculation is most probably due to the viruses being able to replicate only in specific tissue, such as the phloem, within infected plants and it may be difficult to physically insert virions or viral DNA into this tissue. Although in time these difficulties may be overcome by using *Agrobacterium*-mediated transfer (Chapter 6) or by working with protoplasts (see later) they have nevertheless impeded the study of some of the gemini viruses and in some cases precluded the opportunity to investigate the infectivity of DNA

cloned from purified virions. Even so because of their potential as vectors a great deal of attention has been paid to those viruses which can infect monocotyledonous crops such as MSV and WDV.

Replication of Gemini Viruses
Assembly of virion particles takes place in the nucleus of the infected plant cell and double-stranded circular DNAs have been detected which are thought to represent replicative intermediates. Although the genome of the gemini virus is single-stranded DNA (ssDNA), in MSV, a short, approximately 80 base, sequence is found base-paired to the genome. This DNA has ribonucleotides covalently attached at its 5' end, is complementary to a region 5' of two hairpin structures on the MSV genome and is thought to serve as a primer for second strand DNA synthesis upon infection. Although it is considered that second strand synthesis is a prerequisite of viral replication, the presence of a similar primer sequence has not been detected in CLV. Concatomers of viral DNA have been found in infected tissue suggesting that the DNA may replicate by a rolling circle intermediate. In addition variable amounts of concatenated ssDNA, the majority of which are non-covalently associated dimers of the genomic ssDNA, are also found associated with the virion particle (Harrison 1985; Stanley 1985). In addition to the other forms of DNA encapsidated within the gemini virus particle, a discrete subgenomic ssDNA has also been observed in the virions of CLV and TGMV. These are likely to arise as a result of intramolecular recombination taking place during replication.

The gemini viruses are not seed transmitted and as with CaMV there is no evidence to suggest that they integrate into the genome of the host plant.

Genome Organization and Transcription of Gemini Viruses with a Bipartite Genome
The two genomes of the bipartite gemini viruses are referred to as DNA 1 (or A) and DNA 2 (or B) and a number have now been sequenced. In general, the sequence organization of viruses with bipartite genomes are very similar. In each there is an intergenic region of approximately 200 nucleotides common to both DNA components. Although the sequence of the common region varies between different viruses there is a region of approximately 34 nucleotides which can form a stem and loop structure present in each. The role of the common region is not known but it could be important in DNA replication and/or viral packaging. There are six ORFs encoding polypeptides of greater than 10 000 daltons of similar position, size and orientation in CLV, TGMV and BGMV. The ORFs are arrayed on both DNA components in both the viral (+ strand) or complementary strand (− strand) and the ORFs apparently diverge from the common region of the genome (Fig. 4.6). In the complementary strand, all three reading frames are used and the ORFs overlap. Each open reading frame is flanked by sequences similar to those seen flanking other eucaryotic genes and are considered to be sequences controlling transcription and in some cases these may be shared by different ORFs. The function of the proteins encoded by the ORFs largely remains unknown although the comparison of the amino acid sequence of the CLV coat protein with the nucleic acid sequence suggests that the ORF encoding the 30 200 molecular-weight polypeptide is the coat protein gene. There have

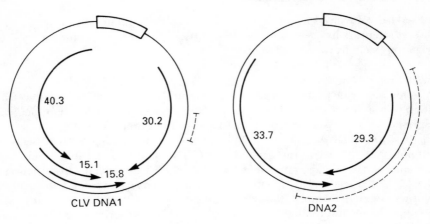

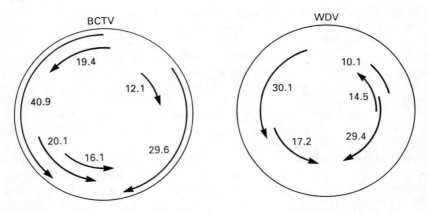

Fig. 4.6 The organization of the gemini virus genome. The thick arrows refer to ORFs encoding polypeptides of molecular weight greater than 10 000 with the numbers referring to the size in kilodaltons (not to scale). In the case of CLV, the ORFs depicted are those that have counterparts in TGMV. The boxed area represents the common region of sequence and the broken lines depict the areas of deletion that have been observed to arise *in vivo*.

been no reports of a systematic study of the gemini virus genome using *in vitro* mutagenesis to investigate the functions of the polypeptides encoded by particular ORFs and which, if any, could be modified by the insertion of foreign DNA. Nevertheless, some experimental observations and analysis of mutants obtained by recombination *in vivo* can be used to draw some conclusions to the possible functions of regions of the viral genomes. With CLV, whereas both cloned components are required to infect whole plants, DNA 1 has been found to be sufficient to initiate replication in protoplasts (Townsend *et al.*, 1986). This would suggest that DNA 1 encodes the viral replicative functions and that DNA 2 is involved in the systemic spread of the virus in whole plants. DNA 2 is also apparently dispensable in leaf hopper transmitted viruses which suggests that it might also encode a factor determining whitefly transmission. The high degree of homology of the coat protein sequence amongst gemini viruses that are transmitted by the same vector suggests that the coat protein also plays an important role in insect transmission. An infectious mutant of CLV has been obtained which contains a deletion in the coat protein gene on DNA 1. This indicates that virion packaging is not a prerequisite for the systemic spread of the virus in the host plant and that the coat protein does not play a role in viral DNA replication. During normal infection by CLV and TGMV deleted forms of DNA 2 accumulate and are packaged within the virion. In order to be maintained the deleted molecule must contain elements required for autonomous replication as well as virion packaging. The subgenomic DNA is approximately half the size of a normal DNA 2 and the deletions encompass a defined region of DNA 2, removing one ORF and the C-terminal sequence of the other (Fig. 4.6). The common region is retained suggesting that it acts in replication and/or packaging of the virion DNA. Double-stranded molecules of a similar size to the deleted molecules are also observed in infected tissue and those that have been studied in detail apparently have a defined region deleted suggesting that intramolecular recombination may take place at a specific site within DNA 2 (MacDowell *et al.*, 1985; Stanley and Townsend 1985). It is probable that intramolecular recombination may also take place in DNA 1 or over other regions of DNA 2, but in these cases the DNA is not maintained nor packaged and hence not observed. In addition, the orientation of the ORFs of the virus genome suggests that transcriptional controls might be present in the intergenic region (Stanley and Townsend 1986).

Analysis of poly(A) RNA isolated from tissue infected with CLV has revealed five viral-specific transcripts hybridizing to both the viral and complementary strands of DNAs 1 and 2. This suggests bidirectional transcription of both DNAs. Each of the ORFs has an RNA associated with them and hybrid-release translation has demonstrated that a 1.0 kb transcript is the coat protein mRNA (Townsend *et al.*, 1985).

Structure of the Gemini Viruses with a Monopartite DNA Genome
The sequence of the monopartite genome of the gemini virus that infects dicotyledonous plants, BCTV, reveals six ORFs encoding polypeptides of molecular weight greater than 10 000. Four resemble those found in the genomes of the whitefly transmitted viruses having a similar size, position and orientation.

However, the sequence of the putative coat protein, the 29 600 molecular weight polypeptide, is more closely related to the leaf hopper transmitted gemini viruses. There is an intergenic region with a hairpin structure similar to that found in double-component viruses (Stanley *et al.*, 1986).

A similar picture is found with the organization of those other viruses transmitted by leafhopper which infect monocotyledonous plants, MSV and WDV. There are seven and five ORFs, respectively, encoding polypeptides of molecular weight greater than 10 000. The ORFs present in WDV have counterparts on the MSV genome. The proteins encoded by MSV and WDV genomes resemble those encoded by DNA 1 of CLV and TGMV. Once again due to the orientation of all but one of the ORFs, transcription is apparently bidirectional from an intergenic region. The intergenic region itself contains sequences capable of forming stem and loop structures as found in the other gemini viruses (MacDowell *et al.*, 1985).

Infectivity of the Cloned Genomes of Gemini Viruses

When the genomes of the gemini viruses are cloned the double-stranded linear viral DNA is infectious. In gemini viruses with a bipartite genome, both cloned components are required to produce infection. Infection takes place at high frequency when the viral sequences are excised from the bacterial cloning vector. But in CLV, infection can result at low frequency from the inoculation of test plants with a mixture of one component excised from the cloning vector and the other component as an uncut plasmid or linearized within the cloning vector. Here, intramolecular recombination, reminiscent of that seen with CaMV, removes most, if not all, of the bacterial sequences from the viral DNA and thus allows infection to take place. In the monopartite genome viruses the cloned BCTV genome is infectious once it has been excised from the plasmid vector; however, infection with cloned MSV DNA has only been obtained with the MSV genome following *Agrobacterium*-mediated transfer of dimers of the genome (Chapter 7).

Gemini Viruses as Vectors

To date there have been no detailed reports of an engineered gemini virus genome being used to transfer foreign genes into plant cells. As described above the deletions that occur within the viral genome during infection might result in the instability of the foreign DNA inserted into the viral genome. Indeed a preliminary report of the insertion of a NPTII gene into a deleted TGMV DNA 2 suggested that it was not stably maintained in inoculated tissue: however, it is not clear whether this was due to rearrangements of DNA or to limitations on viral packaging (MacDowell *et al.*, 1986).

On the other hand the finding that cloned CLV can be used to transfect isolated leaf protoplasts is of great importance because this should allow the development of a transient expression system based on the inoculation of protoplasts with derivatives of the viral genome (Townsend *et al.*, 1986). However, the replication of viral DNA occurs at the same time as cell division suggesting that the two might be linked with viral replication, being cell-cycle dependent.

Conclusions

It is clear that vectors based on viral genomes are unlikely to be as versatile as those based on the Ti plasmid, particularly when used to transform whole plants. Nevertheless one of the major advantages of such vectors, as described at the beginning of this chapter, that of the high levels of expression of foreign DNA in plant cells, has been borne out by the examples of vectors based on the genomes of TMV, BMV and CaMV.

Although the use of vectors based on the genomes of RNA or gemini viruses may not be prone to packaging constraints, in contrast to CaMV, potential difficulties as a result genetic rearrangement remains. Moreover, the accumulation of mutant molecules within the plant cell may interfere with the replication of the non-rearranged molecules. This may be avoided, or at least reduced, if it proves to be possible to select in the whole plant for the presence of non-rearranged molecules containing foreign DNA. At best, these vectors may be able to produce large amounts of a particular protein in all the tissues of an infected plant but will probably lack the precision required to modify the plant phenotype such as may be required in changing a biochemical pathway or a plant's response to a particular stress. In addition the plant may suffer effects of the symptoms produced by a particular virus or be induced to synthesize the—as yet poorly understood—pathogen response polypeptides. Moreover, there is no indication of what the response of the plant will be to making large amounts of a polypeptide for which it has no use.

These difficulties do not arise if viral-based vectors are used in transient expression in isolated cells. In this case vectors based on viral genomes are likely to become useful because of their ability to replicate. These systems, as will be described later, will become increasingly important in testing gene constructions prior to stable introduction into the plant genome. As we have seen transfection of protoplasts with either viral RNA or DNA can be carried out efficiently. This efficiency can be increased further if the chimeric nucleic acid can be packaged *in vitro* and the resultant pseudo-virus particles used to infect protoplasts. This is possible with TMV where an origin of assembly sequence (OAS) in the nucleic acid can direct the assembly of virion particles in the presence of the coat protein. RNAs synthesized *in vitro* bearing the OAS in the correct orientation and a variety of foreign sequences have been found to be packaged *in vitro* (Sleat *et al.*, 1986) and the system has been used to express CAT in tobacco cells (Gallie *et al.*, 1987).

The only limitation with an experimental system based on the expression of foreign DNA in protoplasts is the biochemical status of the isolated plant cell and how this might effect not only the replication of sequences based on viral genomes but also the expression of the inserted gene. For example, the replication of the CLV genome may be dependent on cell cycle and this will effect the levels of vector DNA in the cells. Should the foreign DNA be under the control of a constitutive viral promoter it is likely to be expressed regardless of the state of the transfected cell but if it is under the control of a developmentally regulated promoter it may not be. Very little is known about the expression of individual genes in isolated plant cells cultured under different conditions. This is likely to become the limiting factor in using protoplasts in transient expression.

References

Brisson, N., Paszkowski, J., Penswick, J., Gronenborn, B., Potrykus, I., and Hohn, T. (1984). 'Expression of a bacterial gene in plant cells using a viral vector', *Nature* **310**, pp. 511–514.

Covey, S.N. and Hull, R. (1985). 'Advances in cauliflower mosaic virus research', *Oxford Surveys of Plant Molecular and Cellular Biology*', **2**, pp. 339–346.

Cress, D., Kiefer, M.C. and Owens, R. (1983). 'Construction of infectious potato spindle tuber viroid cDNA clones', *Nucleic Acids Res.* **11**, pp. 6821–6835.

Dougherty, W.G. and Hiebert, E. (1986). 'Genome structure and gene expression of plant RNA viruses', in *Molecular Plant Virology*, vol. II. Ed. Davies, J., pp. 23–81. CRC Press, Boca Raton, Florida.

French, R., Janda, M. and Ahlquist, P. (1986). 'Bacterial gene inserted in an engineered RNA virus: efficient expression in monocotyledonous plant cells', *Science* **231**, pp. 1294–1296.

Gallie, D.R., Sleat, D.E., Watts, J.W., Turner, P.C. and Wilson, T.M.A. (1987). '*In vivo* uncoating and efficient expression of foreign mRNAs packaged in TMV-like particles', *Science* **236**, pp. 1122–1124.

Gronenborn, B., Gardner, R.C., Schaefer, S. and Shepherd, R.J. (1981). 'Propagation of foreign DNA in plants using cauliflower mosaic virus as a vector', *Nature* **294**, pp. 773–776.

Harrison, B.D. (1985). 'Advances in gemini virus research', *Ann. Rev. Phytopath.* **23**, pp. 55–82.

Hull, R. and Covey, S.N. (1983). 'Replication of Cauliflower Mosaic Virus', *Sci. Prog.* (*Oxford*) **68**, pp. 403–422.

Hull, R., Sadler, J. and Longstaff, M. (1986). 'The sequence of carnation etched ring virus DNA: comparison with cauliflower mosaic virus and retroviruses', *EMBO J.* **5**, pp. 3082–3090.

Lefebvre, D.D., Miki, B.L. and Laliberte, J-F. (1987). 'Mammalian metallothionein functions in plants', *Bio/Technology* **5**, pp. 1053–1056.

MacDowell, S.W., MacDonald, H., Hamilton, W.D.O., Coutts, R.H.A. and Buck, K.W. (1985). 'The nucleotide sequence of cloned wheat dwarf virus DNA', *EMBO J.* **4**, pp. 2173–2180.

MacDowell, S.W., Coutts, R.H.A. and Buck, K.W. (1986). 'Molecular characterisation of subgenomic single stranded and double stranded DNA forms from plants infected with tomato golden mosaic virus', *Nucleic Acids Res.* **14**, pp. 7967–7984.

Richins, R.D., Scholthof, H.B. and Shepherd, R.J. (1987). 'Sequence of Figwort Mosaic Virus DNA (caulimovirus group)', *Nucleic Acids Res.* **15**, pp. 8451–8466.

Sleat, D.B., Turner, P.C., Finch, J.J., Butler, J.G. and Wilson, T.M.A. (1986). 'Packaging of recombinant DNA molecules into pseudovirus particles directed by the origin of assembly sequence from tobacco mosaic virus RNA', *Virology* **155**, pp. 299–308.

Stanley, J. (1985). 'Gemini viruses', *Advances in Virus Research* **30**, pp. 139–177.

Stanley, J. and Townsend, R. (1985). 'Characterisation of DNA forms associated with cassava latent virus infection', *Nucleic Acids Res.* **13**, 2189–2206.

Stanley, J. and Townsend, R. (1986). 'Infectious mutants of cassava latent virus generated *in vivo* from intact recombinant DNA clones containing single copies of the genome', *Nucleic Acids Res.* **14**, pp. 5981–5998.

Stanley, J., Markham, P.G., Callis, R.J. and Pinner, M.S. (1986). 'The nucleotide sequence of an infectious clone of beet curly top virus', *EMBO J.* **5**, pp. 1761–1767.

Takamatsu, N., Ishikawa, M., Meshi, T. and Okada, Y. (1987). 'Expression of bacterial chloramphenicol acetyltransferase gene in tobacco plants mediated by TMV-RNA', *EMBO J.* **15**, pp. 307–311.

Thomas, C.M., Hull, R., Bryant, J.A. and Maule, A.J. (1985). 'Isolation of a fraction from cauliflower mosaic virus infected protoplasts which is active in the synthesis of (+) and (−) stranded viral DNA and reverse transcription of primed RNA templates', *Nucleic Acids Res.* **13**, pp. 4557–4575.

Townsend, R., Stanley, J., Curson, S.J. and Short, M.N. (1985). Major polyadenylated transcripts of cassava latent virus and location of the gene encoding coat protein', *EMBO J.* **4**, pp. 33–37.

Townsend, R., Watts, J. and Stanley, J. (1986). 'Synthesis of viral DNA forms in *Nicotiana plumbaginifolia* protoplasts inoculated with cassava latent virus (CLV); evidence for the independent replication of one component of the CLV genome', *Nucleic Acids Res.* **14**, pp. 1253–1265.

Walden, R. and Howell, S.H. (1982). 'Intergenomic recombination event among pairs of defective cauliflower mosaic virus genomes', *J. Mol. Appl. Genet.* **1**, pp. 447–456.

Chapter 5

Naked DNA Transformation of Plant Cells

Potential Advantages of Transformation with Naked DNA

As described in Chapter 1, the first experiments involving the attempted transfer of foreign DNA to plant cells took place in the early 1970s and centred on the uptake of DNA by either whole plants, suspension cells or isolated protoplasts. However, with doubt cast on the success of such experiments, attention shifted to the development of vectors based on either the Ti plasmid or plant viruses. Nevertheless, as work has progressed with these experimental systems, interest returned to the uptake of naked DNA by plant cells because, as discussed in Chapter 4, there are inherent experimental limitations in both the Ti plasmid and viral transformation systems. For example, if naked DNA uptake was applicable to a wide variety of species, it may be used to transform those plants which are, as yet, not amenable to *Agrobacterium*-mediated gene transfer. Moreover, naked DNA uptake might offer a method by which a large number of protoplasts could take up multiple copies of DNA, allowing the study of not only transient gene expression but also the functional activity of specific sequences of DNA, such as origins of replication.

Generally, the approaches adopted for the transformation of plant cells with naked DNA have been those which have been used successfully with animal or yeast cells. The difficulties faced in such experiments lay in the survival of the plant tissue following a particular treatment and demonstrating that the DNA had entered the plant cell and was being expressed. The original experiments in the area of naked DNA uptake were unable to conclusively demonstrate that transformation had taken place principally because of a lack of suitable genetic markers and, prior to the development of nucleic acid hybridization techniques, an inability to distinguish between DNA stably inserted into the plant cell and contaminating bacterial or inoculated DNA. Clearly, to show definitively that a plant or plant cell is transformed, a new phenotype and the presence of foreign DNA being expressed must be demonstrated. In whole plants the new phenotype must be inherited in a stable manner. This requires the use of genetic markers which are applicable to plant systems, for example either the correction of auxotrophy or, preferably, dominant selection, where activity can be unambiguously linked to the presence of foreign DNA in the plant cell.

Initial Success with Transformation of Plants with Naked DNA

The first experiments involving the transformation of plant cells with naked DNA conforming to these criteria were carried out using Ti plasmid DNA to transform *Petunia* (Davey *et al.*, 1980) or tobacco cells (Krens *et al.*, 1982). In the latter case, protoplasts were incubated with Ti plasmid DNA in the presence of calcium ions and polyethylene glycol (PEG). It is not clear how PEG (or for that matter, other long-chain cations such as poly-L-ornithine) act to induce DNA uptake, although in protoplast fusion PEG is used to minimize charge repulsion. It is also thought to remove 'free water' from a solution thus precipitating large molecules, such as DNA, as well as inducing endocytosis by the protoplasts. Following PEG treatment the protoplasts were cultured in the presence of phytohormones to induce cell division and callus formation with the resulting calli being transferred to hormone-free media. Putative transformants, selected for by their ability to grow in the absence of phytohormones, were tested for octopine synthase activity. A variety of transformants were obtained and in some cases, where callus could form shoots, these were unable to form roots and lacked apical dominance, characteristics common in transgenic plants obtained from cells which had been transformed by *Agrobacterium*. Southern blot analysis was used to demonstrate the presence of Ti plasmid DNA stably inserted into the plant nuclear genome. However, it was found that the Ti plasmid DNA that had integrated into the genome was not uniform in size between different transformants and that rearrangements, multiplications and scrambling of the DNA had occurred suggesting that the process of integration was different to that which takes place during transformation by *Agrobacterium*. Nevertheless, these were the first reports of the stable transformation of a plant cell by naked DNA and demonstrated that tumour formation was dependent solely on the Ti plasmid; however, they did not exclude the possibility that Ti plasmid-specific factors were essential for transformation.

Transformation of *Chlamydomonas*

At the same time, the transformation of the green algae *Chlamydomonas reinhardii* was reported (Rochaix and van Dillewijn 1982). In these experiments an *arg7* cell wall-deficient strain of *C. reinhardii* was used. The *arg7* locus encodes arginosuccinate lyase, the last enzyme of the arginine biosynthetic pathway. The DNA that was used in the transformation was the plasmid, pYe*arg4*, which contains a yeast origin of replication and the yeast *arg4* locus which is analogous to the *C. reinhardii arg7* locus. Two transformation methods were used, one involving incubation of the cells with the DNA in the presence of poly-L-ornithine, the other, which is similar to a method used for yeast transformation, involved incubating the cells with the DNA in the presence of PEG. In both cases *arg*$^+$ cells were obtained and these contained a portion of the yeast plasmid integrated in the nuclear DNA. Although the *arg*$^+$ phenotype was stable following further growth under non-selective conditions, the yeast DNA within the nucleus appeared to become smaller suggesting that rearrangements of the DNA had taken place during culture of the cells. Subsequently, vectors were developed that could replicate autonomously in *Chlamydomonas* (Rochaix *et al.*, 1984). These consisted of the yeast *arg4* locus cloned in pBR322 and random fragments of *Chlamydomonas* DNA. Four fragments, able to confer autonomous replication, were studied in detail, and were found to have been derived from the chloroplast genome; one fragment encompassed a region thought to contain the chloroplast origin of replication. The plasmids replicate in the cells and can be rescued by isolating total DNA from the *Chlamydomonas* and using it to transform *E. coli*. However, with prolonged culture the copy number of the plasmid within the cells appears to decline. It is not known if this decline is due to failure of the plasmids to segregate correctly or reversion of the *arg7* mutation removing the selective pressure.

Unfortunately this system suffers from a very low transformation frequency (between 1×10^{-6} and 1×10^{-7}) and the cell wall-deficient mutants are extremely fragile. More recently, improved frequencies of transformation (1×10^{-6}) have been reported following transformation using a plasmid containing a NPTII gene linked to an SV40 promoter and a yeast 2 μm plasmid origin of replication. The resulting transformants were resistant to kanamycin and appeared to maintain the plasmid in the cytoplasm (Hashnan *et al.*, 1985). This work demonstrates the feasibility of using dominant selection with *Chlamydomonas* and suggests that the yeast 2 μm plasmid origin of replication is functional in *Chlamydomonas* as is the SV40 promoter/enhancer sequence. A great number of nuclear mutants of *Chlamydomonas* have been obtained affecting a wide range of functions including photosynthesis and the cell cycle. An efficient transformation system for *Chlamydomonas* would allow the rescue of genes encoding these functions from other plant sources by complementation.

Plant Transformation by the Uptake of a Dominant Selectable Marker Gene

As we have already seen, central to the development of plant transformation vectors has been the construction of dominant selectable marker genes functional in plant cells. These can be used to demonstrate that transformation has taken place by their ability to confer a specific, selectable phenotypic change on the plant cell. This can be confirmed if the marker directs the synthesis of a unique enzyme that can be simply assayed and the physical presence of the foreign DNA in the transformant can be demonstrated by Southern blot analysis. Using such a marker allowed the first demonstration of direct gene transfer to higher plant cells by Paszkowski and co-workers (1984). Here a plasmid, based on the bacterial cloning vector pUC9, was constructed to contain the NPTII gene of Tn 5 fused in frame with the first 23 codons of ORF VI of CaMV flanked by the 19S RNA promoter and polyadenylation sequences. Protoplasts were incubated with this DNA under conditions similar to those used previously by Krens and co-workers and induced to divide and produce callus. Subsequent growth and selection of transformants was greatly facilitated by the development in the same laboratory of the 'bead culture' technique (Shillito *et al.*, 1983). This involves embedding the microcalli in agarose discs which are incubated in liquid media, allowing replacement of the culture media so that the selective antibiotic can be added and replaced at regular intervals, thus maintaining selective pressure during the first 4 weeks of culture. After this time resistant colonies become visible against a background of dead or dying sensitive colonies (Fig. 5.1). The resistant colonies can be induced to form shoots and the resulting plantlets grow normally in the presence of kanamycin. Moreover, explants from different regions of the resistant plant produce resistant cell cultures indicating that no loss of the resistance phenotype occurred at any particular stage of development. In addition the resistant tissue contained NPTII enzyme activity. Analysis of the growth of microspores and genetic crossing indicated that resistance was maintained to a level of approximately 50% in the male germline and could be inherited as a single dominant trait. Southern blot analysis indicated that 3–5 copies of the NPTII gene had integrated into the high molecular weight nuclear DNA. The transformation frequencies originally obtained by this technique were 1×10^{-5}, although modifications of the technique have been carried out in a number of laboratories and frequencies of up to 4.8% of the surviving clones have been reported as being transgenic (Negrutiu *et al.*, 1987). Subsequently it has been found that co-transformation of two unlinked genes can take place at high frequency following naked DNA uptake. This allows a rapid and simple method of introducing into the plant genome genes for which there is no direct selection (Schöcher *et al.*, 1986).

Following the uptake of DNA by the plant cells the mechanism by which the DNA integrates into the nuclear genome is not known, but as seen earlier with the direct uptake of the Ti plasmid, it appears to differ from that which occur in transformation mediated by *Agrobacterium*. Comparison of the organization of the foreign DNA integrated into the nuclear genome in transgenic tissue resulting from naked DNA uptake with that seen after co-cultivation with *Agrobacterium*, indicates that transgenic cells resulting from the former process contain a complex array of concatenated DNA whereas those resulting from the latter process contain no extensive modifications of the transferred DNA except for formation of tandem dimers (Hain *et al.*, 1985). Although the foreign DNA may be present in high copy number it does not appear that these sequences are distributed at random but are generally located at one or two sites in the genome (Potrykus *et al.*, 1985a). It is difficult to assess whether the rearrangements observed following

(A)

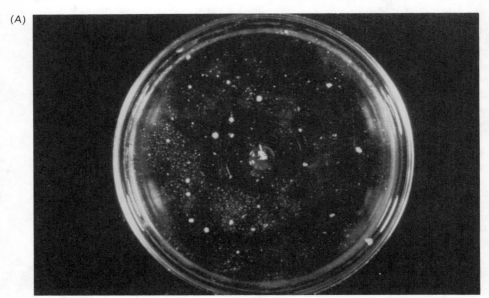

(B)

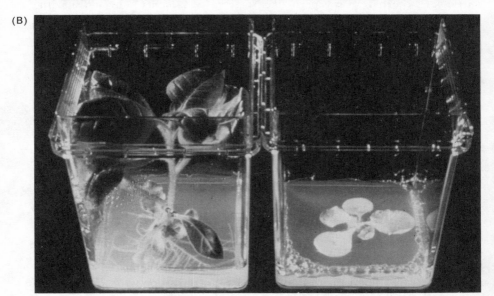

Fig. 5.1 Resistant cell colonies and plants following naked DNA uptake.
(A). Tobacco protoplasts were incubated in the presence of a chimeric NPT
II gene and resistant calli are seen growing against a background of dying
cells under selection. (B). Wild type and transformed shoots growing in the
presence of kanamycin. (Reproduced with permission from Paszkowski *et al.*,
1984).

naked DNA uptake occur prior to or following integration of the DNA into the plant genome. It is thought, however, that the DNA may form concatemers or be modified shortly after its entrance into the plant cell and that these altered molecules are multiplied, by an unknown process, prior to integration. Detailed analysis of the foreign DNA integrated into the genome following naked DNA uptake suggests that the integrated DNA occurs in multiple copy arrays of tandem sequences mostly comprising identically altered units of donor DNA. Between 10 and 25 copies of DNA may be joined in this manner. Because the DNA that is integrated into the genome appears to be stable during callus and plant development, as well as meiosis, it is thought that extensive alteration to the DNA by homologous recombination after integration does not take place. To date there does not appear to be any evidence to suggest that increased copy number of inserts can confer resistance to increased levels of the selectable marker and there appears to be no correlation between copy number and expression level. Moreover, the DNA is apparently stable when the tissue is incubated in the absence of selection (Czernilofsky *et al.*, 1986).

Similar experiments have been successfully carried out with turnip protoplasts (Paszkowski *et al.*, 1986) and with cells derived from suspension cultures of wheat (*Triticum monococcum*) (Lorz *et al.*, 1985) Italian ryegrass (*Lolium multiflorum*) (Potrykus *et al.*, 1985b) and sugar-cane (Chen *et al.*, 1987). The latter experiments have demonstrated that naked DNA uptake could be used to transfer DNA into monocotyledonous cells and that the dominant selectable markers which had been developed for dicotyledonous cells, using the *nos*, 35S and 19S CaMV RNA promoters, were also functional in these systems. Hence it is likely that naked DNA uptake will be applicable to any plant species from which cells can be obtained. However, the major limitation of the system remains in plant regeneration. Whereas plant regeneration from monocotyledonous species may be possible from partially differentiated tissue, such as somatic embryos, protoplasts derived from suspension cell culture can, in general, only be regenerated to the callus stage, although recently the regeneration of rice (Abdullah *et al.*, 1986) and maize (Rhodes *et al.*, 1987) has been reported.

Transient Gene Expression in Protoplasts

The experiments that have been described so far have concerned themselves with the stable transformation of the plant cell where gene expression in the transgenic tissue is analysed weeks, if not months, after the transfer of the foreign DNA into the cell. However, if the method of DNA uptake is efficient and a large proportion of cells take up several copies of DNA sequence, then transient expression of the DNA, i.e. expression of the DNA as soon as it enters the plant cell, may be detectable. Such an experimental system would be extremely useful in screening a large number of gene constructs which could be introduced separately into protoplasts and assayed for expression days, and not weeks, after transfection. Transient expression systems rely on the ability to efficiently introduce DNA (or RNA) into isolated cells, a sensitive method for detecting gene expression, and a gene construct which is expressed in the cell under the culture conditions used in the experiment.

Electroporation

Although transient gene expression can be observed in cells following DNA uptake in the presence of PEG, it has been studied most extensively following electroporation of plant cells. Electroporation relies on electrical impulses of a high field strength reversibly permeabilizing the cell membrane allowing the entrance of DNA. This is generally considered to be an extremely efficient way of introducing foreign nucleic acids into plant cells. The two most important factors affecting the process are the strength of the electric field, which depends on the diameter of the protoplast, and the pulse decay time. Protoplasts are suspended in a buffered saline solution containing plasmid DNA in a cuvette and a pulse, for example, of 200 V for 54 msec is applied by the discharge of a capacitor across two platinum electrodes in the cuvette. Initial work used two different plasmids based on pBR322. One, pNOSCAT, contained a CAT gene flanked by the *nos* promoter and poly(A) addition site whereas the other had the *nos* promoter replaced by the 35S RNA promoter from CaMV. The protoplasts were cultured and CAT activity was detectable 3–96 h after electroporation with peak activity occurring between 24 and 48 h. This technique was initially used to demonstrate transient expression in carrot, tobacco and maize cells (Fromm *et al.*, 1985). Similar results have been obtained with protoplasts from rice leaves and callus (*Oryza sativa*), wheat suspension cells (*Triticum monococcum*), sorghum callus (*Sorghum bicolor*) (Ou-Lee *et al.*, 1986) and a variety of other plant cells (Hauptmann *et al.*, 1987). Transient expression of mRNA electroporated into cells has also been obtained (Callis *et al.*, 1987).

Electroporation has also been used to obtain stable transformants of tobacco (Shillito *et al.*, 1985) and a cell line of maize (*Zea mays* cv. 'Black Mexican') (Fromm *et al.*, 1986). Transgenic tobacco plants which have been obtained from electroporated protoplasts display a complex pattern of integrated DNA similar to that following PEG treatment. Moreover, the DNA appeared to have undergone rearrangements prior to integration with plasmid recircularization and the formation of concatemers (Riggs *et al.*, 1986).

Effect of Antisense RNA on Transient Expression

Transient gene expression has been used to investigate the possible inhibition of gene expression by the synthesis of antisense RNA (Ecker and Davies 1986). In this case electroporation was carried out with two different plasmid constructs, one containing a promoter linked to the CAT gene in the normal orientation and the other having the promoter linked to the CAT gene in the opposite, antisense, orientation. When the proportion of the sense to antisense DNA electroporated into cells was 1:100, CAT activity was reduced by more than 95%. Using different promoter sequences, it was demonstrated that the level of inhibition is correlated with promoter strength suggesting that inhibition was not simply a result of the sense and antisense plasmid DNA hybridizing to each other. The mechanism by which inhibition takes place is not known but it is thought that the hybridization

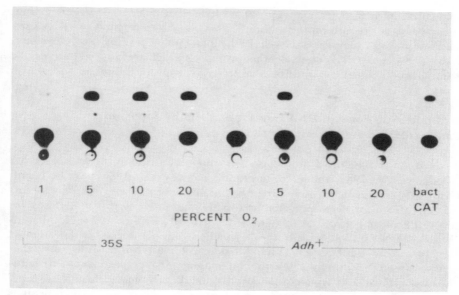

Fig. 5.2 Effect of O_2 tension on expression of CAT constructs. Constructs containing either the CaMV 35S RNA or ADH promoters upstream from the CAT gene were electroporated into maize protoplasts which were subsequently incubated under different oxygen tensions as shown and assayed for CAT activity. Except for low O_2 tension, the 35S RNA promoter is insensitive to O_2 tension whereas the ADH promoter reaches optimal activity at 5% O_2. (Reproduced with permission from Howard *et al.*, 1987.)

to each other of sense and antisense RNA may block passage of the mRNA to the cytoplasm, enhance turnover of the RNA or interfere with the translation of the RNA.

Transient Expression as a Tool to Analyse Promoter Sequences
Another example of the use of transient expression in studying the control of gene expression has been provided in studies concerning alcohol dehydrogenase (ADH). ADH is induced in response to anaerobiosis and it has been found that a chimeric gene consisting of the maize ADH promoter fused to the CAT gene and the *nos* poly(A) addition site can be expressed in protoplasts under low oxygen tension following electroporation into cells derived from a 'Black Mexican' maize cell line (Fig. 5.2). The optimum oxygen tension for induction is the same for both the chimeric construct and the endogenous ADH gene and induction is regulated at the level of mRNA, transcription being initiated at the same site in both cases (Howard *et al.*, 1987). This type of experiment has allowed the identification of the region upstream from the pea ADH gene thought to be important in the *cis*

regulation of anaerobic gene expression (Llewellyn *et al.*, 1987). This experimental system depends on the ability of isolated cells to respond to low oxygen levels in the same manner as the cells in the intact plant. If transient expression is to be used as a tool in the study of other promoters which are induced by a specific stimulus the isolated cell must be able to respond to that stimulus.

Transferring Foreign DNA into Plant Cells by Liposome Encapsidation and Microinjection

Both fusion of mesophyll protoplasts with liposomes containing plasmid DNA (Deshayes *et al.*, 1985) and microinjection (Crossway *et al.*, 1986) have been used to produce transgenic plants but these techniques have not been as widely used as PEG treatment or electroporation presumably because experimentally they are more difficult to perform. Although used extensively with animal cells, microinjection of DNA into plant cells has achieved only limited success to date. This is largely because of difficulties in immobilizing protoplasts prior to injection and in injecting the DNA into the cell without damaging the tonoplast which surrounds the plant cell vacuole (which can, depending on the cell type, make up between 5 and 95% of the total cell volume). In an attempt to overcome these difficulties Crossway *et al.*, (1986) have developed the 'holding pipette' technique to microinject tobacco protoplasts. This involves the protoplasts being held onto a 5–10 μm pipette by gentle suction. The protoplasts can be turned so that the nucleus, seen under differential optics, is accessible to injection by a 0.2 μm diameter injection pipette. Using such an arrangement 2 picolitres of DNA solution can be injected either into the cytoplasm or the nucleus. Following microinjection the protoplasts are cultured in droplets of media on the reverse side of a Petri dish through which the growth of microcalli can be observed. After approximately 2 months, the calli are transferred to solid media and cultured for a further 4 months in order to produce enough material for analysis. The transformation efficiency resulting from microinjecting nuclei was 14%, approximately twice that obtained from injecting the DNA into the cytoplasm. As found with the other techniques of naked DNA uptake the site of integration appeared to be random but interestingly, in contrast, single copy insertions took place. The success of microinjection applied to plant cells, though technically difficult when compared to some of the other techniques, raises the possibility of microinjecting a variety of materials such as chromosomes and it has been suggested that chloroplasts and mitochondria could be transferred by microinjection.

Conclusions

It is obvious from the experiments described in this chapter that naked DNA can be introduced into protoplasts isolated from a variety of plant species, both monocotyledonous and dicotyledonous. The techniques that have been adopted can overcome the apparent host range limitations inherent with *Agrobacterium*-mediated gene transfer. However, the difficulty in achieving plant regeneration from single cells from many plant species, particularly monocotyledonous ones, remains. This may, in time, be overcome but currently the regeneration of plants from many agronomically important species can only be achieved from partially differentiated tissue. Hence, in the absence of plant regeneration from single cells, methods need to be devised by which DNA can be introduced directly into morphogenic tissue which can regenerate *in vitro*. While this has been achieved with the transformation of pro-embryogenic suspension cells of carrot with *Agrobacterium* this has yet to be demonstrated with naked DNA. Macroinjection might be feasible and recently a method has been devised by which DNA is coated onto 4 μm tungsten particles and shot into plant tissue using a particle gun (Klein *et al.*, 1987). The technique was used to introduce viral RNA and plasmid DNA into intact onion epidermal cells. Three days following bombardment the cells accumulated viral inclusion bodies and CAT activity could be detected. The target onion cells are relatively large but it is thought that the process could be refined so that meristematic, or embryogenic, callus could be bombarded. It will, however, be no easy task to introduce the DNA directly into the meristematic tissue which is likely to be protected by several cell layers.

References

Abdullah, R., Cocking, E.C. and Thompson, J.A. (1986). 'Efficient plant regeneration from rice protoplasts through rice somatic embryogenesis', *Bio/Technology* **4**, pp. 1087–1090.

Callis, J., Fromm, M. and Walbot, V. (1987). 'Expression of mRNA in plant and animal cells', *Nucleic Acids Res.* **15**, pp. 5823–5831.

Chen, W.H., Gartland, K.M., Davey, M.R., Sotak, R., Gartland, J.S., Mulligan, B.J., Power, J.P. and Cocking, E.C. (1987). 'Transformation of sugarcane protoplasts by direct uptake of a selectable chimeric gene', *Plant Cell Reports* **6**, pp. 297–301.

Crossway, A., Oakes, J.V., Irvine, J.M., Ward, B., Knauf, V.C. and Shewmaker, C.K. (1986). 'Integration of foreign DNA following microinjection into tobacco mesophyll protoplasts', *Mol. Gen. Genet.* **303**, pp. 179–185.

Czernilofsky, A.P., Hain, R., Herrerra-Estrella, L., Lorz, H., Goyvaerts, E., Baker, B. and Schell, J. (1986). 'Fate of selectable marker DNA integrated into the genome of *Nicotiana tabacum*', *DNA* **5**, pp. 101–113.

Davey, M.R., Cocking, E.C., Freeman, J., Pearce, N. and Tudor, I. (1980). 'Transformation of petunia protoplasts by isolated *Agrobacterium* plasmid', *Plant Sci. Lett.* **18**, pp. 307–313.

Deshayes, A., Herrera-Estrella, L. and Caboche, M. (1985). 'Liposome-mediated transformation of tobacco mesophyll protoplasts by an *Escherichia coli* plasmid', *EMBO J.* **4**, pp. 2731–2737.

Ecker, J.R. and Davies, R.W. (1986). 'Inhibition of gene expression in plant cells by expression of antisense RNA', *Proc. Nat. Acad. Sci. USA* **83**, pp. 5372–5376.

Fromm, M., Taylor, L.P. and Walbot, V. (1985). 'Expression of genes transferred to monocot and dicot plant cells by electroporation', *Proc. Nat. Acad. Sci. USA* **82**, pp. 5824–5828.

Fromm, M., Taylor, L.P. and Walbot, V. (1986). 'Stable transformation of maize after gene transfer by electroporation', *Nature* **319**, pp. 791–793.

Hain, R., Stabel, P., Czernilofsky, P.A., Steinbiss, H.H., Herrera-Estrella, L. and Schell, J. (1985). 'Uptake, integration, expression and genetic transmission of a selectable chimaeric gene by plant protoplasts', *Mol. Gen. Genet.* **199**, pp. 160–168.

Hashnan, S.E., Manavathu, E.K. and Leung, W.-C. (1985). 'DNA-mediated transformation of *Chlamydomonas reinhardii* cells: use of aminoglycoside 3′-phosphotransferase as a selectable marker', *Mol. Cell. Biol.* **5**, pp. 3647–3650.

Hauptmann, R.M., Ozias-Atkins, R., Vasil, V., Tabaeizadeh, Z., Rogers, S.G., Horsch, R.B., Vasil, I.K. and Fraley, R.T. (1987). 'Transient expression of electroporated DNA in monocotyledonous and dicotyledonous species', *Plant Cell Reports* **6**, pp. 265–270.

Howard, E.A., Walker, J.C., Dennis, E.S. and Peacock, W.J. (1987). 'Regulated expression of an alcohol dehydrogenase 1 chimeric gene introduced into maize protoplasts', *Planta* **170**, pp. 535–540.

Klein, T.M., Wolf, E.D., Wu, R. and Stanford, J.C. (1987). 'High velocity microprojectiles for delivering nucleic acids into living cells', *Nature* **327**, pp. 70–73.

Krens, F.H., Molendijk, L., Wullems, G.J., Schilperoot, R.A. (1982). '*In vitro* transformation of plant protoplasts with Ti plasmid', *Nature* **296**, pp. 72–74.

Llewellyn, D.J., Finnegan, E.J., Ellis, J.G., Dennis, E.S. and Peacock, W.J. (1987). 'Structure and expression of an alcohol dehydrogenase 1 gene from *Pisum sativum* (cv. 'Greenfeast')', *J. Mol. Biol.* **195**, pp. 115–123.

Lorz, H., Baker, B. and Schell, J. (1985). 'Gene transfer to cereal cells mediated by plasmid transformation', *Mol. Gen. Genet.* **199**, pp. 178–182.

Ou-Lee, T.M., Turgeon, R. and Wu, R. (1986). 'Expression of a foreign gene linked to either a plant virus or a *Drosophila* promoter after electroporation of protoplasts of rice, wheat, and sorghum', *Proc. Nat. Acad. Sci. USA* **83**, pp. 6815–6819.

Negrutiu, I., Shillito, R., Potrykus, I., Biasini, G. and Sala, F. (1987). 'Hybrid genes in the analysis of transformation conditions. I. Setting up a simple method for direct gene transfer in plant protoplasts', *Plant Mol. Biol.* **8**, pp. 363–373.

Paszkowski, J. and Saul, M. (1986). 'Direct gene transfer to plants', *Methods in Enzymology* **118**. Eds. Weisbach, A. and Weisbach, H., pp. 668–684. London, Academic Press.

Paszkowski, J., Pisan, B., Shillito, R.D., Hohn, T., Hohn, B. and Protrykus, I. (1986). 'Genetic transformation of *Brassica campestris* var. *rapa* protoplasts with an engineered cauliflower mosaic virus', *Plant Mol. Biol.* **6**, pp. 303–312.

Paszkowski, J., Shillito, R.D., Saul, M., Mandak, V., Hohn, T., Hohn, B. and Potrykus, I. (1984). 'Direct gene transfer to plants', *EMBO J.* **3**, pp. 2717–2722.

Potrykus, I., Paskowski, J., Saul, M., Petruska, J. and Shillito, R.D. (1985a). 'Molecular and general genetics of a hybrid foreign gene introduced into tobacco by direct gene transfer', *Mol. Gen. Genet.* **199**, pp. 169–177.

Potrykus, I., Saul, M., Petruska, J., Paskowski, J. and Shillito, R.D. (1985b). 'Direct gene transfer to cells of a graminaceous monocot', *Mol. Gen. Genet.* **199**, pp. 183–188.

Rhodes, C.A., Lowe, K.S. and Ruby, K.L. (1987). 'Plant regeneration from protoplasts isolated from embryogenic maize cell cultures', *Bio/Technology* **6**, pp. 56–60.

Riggs, C.D. and Bates, C.W. (1986). 'Stable transformation of tobacco by electroporation: Evidence for plasmid concatenation', *Proc. Nat. Acad. Sci. USA* **83**, pp. 5602-5604.

Rochaix, J-D. and Van Dillewijn, J. (1982). 'Transformation of the green alga *Chlamydomonas reinhardii* with yeast DNA', *Nature* **296**, pp. 70–72.

Rochaix, J-D., Van Dillewijn, J. and Rahire, M. (1984). Construction and characterisation of autonomously replicating plasmids in green unicellular alga *Chlamydomonas reinhardii*', *Cell* **36**, pp. 925–931.

Schöcher, R.J., Shillito, R., Saul, M., Paszkowski, J. and Potrykus, I. (1986). 'Co-transformation of unlinked foreign genes into plants by direct gene transfer', *Bio/Technology* **4**, pp. 1093–1096.

Shillito, R.D., Paszkowski, J. and Potrykus, I. (1983). 'Agarose plating and bead type culture technique enable and stimulate development of protoplast-derived colonies in a number of plant species', *Plant Cell Reports* **2**, pp. 244–247.

Shillito, R.D., Paszkowski, J. and Potrykus, I. (1985). 'High efficiency gene transfer to plants', *Bio/Technology* **3**, pp. 1099–1103.

Chapter 6

Gene Expression in Transgenic Tissue

Introduction

As was apparent in Chapters 3 and 5, techniques for the production of transformed plants have become well established and there are now a wide variety of vectors with appropriate genetic markers available for the insertion of foreign DNA into plant cells. Transgenic plants have been obtained from a number of different species (see Table 6.1) and this list is likely to increase as the techniques of transformation, coupled with those of plant regeneration, are applied to more species. However, it should be noted that the regeneration of transformed plants from some of the world's most important crops, such as the graminaceous monocotyledonous species and the majority of the forage legumes, has yet to be reported, although in some cases transgenic cell lines have been obtained by *Agrobacterium*-mediated transformation (Chapter 3) or naked DNA uptake (Chapter 5). Nevertheless, a great deal of information has been obtained from those species from which transgenic plants can be obtained and some of the basic principles concerning the expression of foreign DNA in plants are beginning to become apparent.

Variation in the Levels of Expression of Foreign DNA Sequences

Before describing some of the results obtained concerning the expression of foreign DNA introduced into the plant genome, it is important to note that the levels of

Table 6.1 Transgenic plants produced by *Agrobacterium*-mediated transformation

Plant	Reference
Arabidopsis thaliana	Lloyd *et al.*, 1986
Armoracia lapathifolia (horse-radish)	Noda *et al.*, 1987
Asparagus officinalis	Bytebier *et al.*, 1987
Brassica napus (oilseed rape)	Guerche *et al.*, 1987
Convolvulus arvensis (morning glory)	Tepfer 1984
Cucumis sativus (cucumber)	Trulson *et al.*, 1986
Daucus carota (carrot)	Tepfer 1984
Dioscorea bulbifera (yam)	Schafer *et al.*, 1987
Gossypium hirsutum (cotton)	Umbeck *et al.*, 1987
Helianthus anuus (sunflower)	Everett *et al.*, 1987
Lactuca sativa (lettuce)	Michelmore *et al.*, 1987
Linum usitatissimum (flax)	Basiran *et al.*, 1987
Lotus corniculatus (bird's foot trefoil)	Petit *et al.*, 1987
Lycopersicon esculentum (tomato)	McCormack *et al.*, 1986
Medicago varia (alfalfa)	Deak *et al.*, 1986
Nicotiana tabacum (tobacco)	Zambryski *et al.*, 1984
Nicotiana plumbaginifolia	Horsch *et al.*, 1984
Petunia hybrida	Horsch *et al.*, 1985
Populus NC-5539	Fillatti *et al.*, 1987
Solanun nigrum	Wei *et al.*, 1986
Solanum tuberosum (potato)	Ooms *et al.*, 1987
Stylosanthes humilis	Manners 1987
Trifolium ripens (white clover)	White and Greenwood 1987

expression of the transferred DNA in transgenic tissue can vary greatly between different individual transformants and that the levels of transcription directed by a specific 'foreign' promoter might not be the same as those directed by the endogenous homologous promoter. This is of special importance when the comparison of levels of expression of a number of individual gene constructs in transgenic tissue is required. This variation in the levels of expression of a particular construct between individual transformants appears to be a general phenomenon and has been systematically studied in transgenic *Petunia* engineered to contain the *ocs* gene fused to the promoter of a light-harvesting chlorophyll *a/b* binding protein (*1hcp*) gene derived from *Petunia* (Jones *et.*, 1985). In this study it was found that although the levels of expression could be comparable and qualitatively the same as with the endogenous homologous promoter, there was a greater than 200-fold variation in expression of the foreign sequences between different transformants as judged by the steady-state levels of *ocs*-specific RNA. Moreover, when the expression of the *1hcp* promoter–*ocs* fusion was compared with that of a *nos* promoter–NPTII gene fusion contained on the same construct, there was a great deal of variation in different transformants. The reason for this variation is not understood; for example it cannot be fully accounted for by different copy numbers of the transferred DNA. Generally, because it is considered that the T-DNA inserts at random into the plant genome, it is thought that the variation is due to the local environment into which the foreign DNA is inserted and this has come to be known as 'position effect'. Position effects might result from a number of different factors such as the insertion of the DNA into a region of the plant genome that is not normally transcribed or near sequences that can control transcription in a positive or negative manner, methylation of the inserted DNA or the construct itself having some inherent feature which affects its transcription in its new environment. Hence, in order to be able to assess the expression of a particular gene construct, it is important to minimize the effect of this variation by analysing a large number of individual transformants.

Analysis of Promoter Function

Many of the constructs that have been inserted into plants have used promoters derived from either the Ti plasmid or the CaMV genome to direct expression of foreign genes; gene transfer has allowed the detailed analysis of these promoters in order to assess which sequences are important in directing gene expression. In addition, gene transfer has been used most effectively to dissect the sequences located upstream from developmentally-regulated genes in order to define which are responsible for the tissue-specific and developmentally-regulated transcription of these genes. To date, this has been generally studied with genes whose expression is simple to experimentally manipulate and/or whose products are relatively abundant.

Constitutive Promoters
The *nos* promoter, generally considered to be constitutive in its mode of expression, has been studied in detail. For example, deletions of the *nos* promoter fused to the CAT gene have been inserted into a binary vector and used to assess expression in transformed tobacco callus (An *et al.*, 1986). It was shown that three specific regions of the DNA upstream from the transcriptional start site of the *nos* gene were required for efficient transcriptional activity (Fig. 6.1A). These include sequences which are similar to those found to be important for transcription in animal systems, the CCAAT and TATA boxes, at -78 to -70 and -26 to -19 respectively, from the start of transcription and a region located between -170 and -100 which contains a near perfect 11 bp direct repeat sequence and a 8 bp inverted repeat sequence which may be important in conferring constitutive gene expression on the *nos* promoter.

Another example of a T-DNA promoter construct that can be used to direct expression of foreign DNA is that derived from the dual promoter of genes 1 and 2 of the T_R-DNA of pTiAch5 (see Fig. 2.4, p.13). These genes encode the two most abundant transcripts in transformed tissue and are transcribed divergently, both being initiated from a 476 bp fragment of DNA (Velten *et al.*, 1984). The dual promoter can be linked to two different marker genes which can be expressed in transgenic tissue with the relative transcriptional activities of both promoters

(A)

The nos promoter.

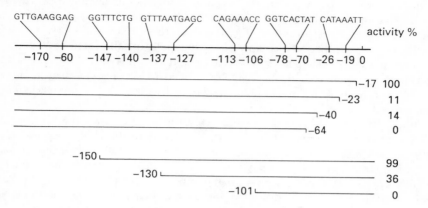

(B)

The CaMV 35S RNA promoter.

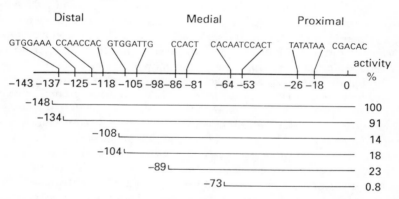

Fig. 6.1 Activity maps of the (A) *nos* promoter and (B) CaMV 35S RNA promoter. Sequences considered important in transcription are indicated with their position with respect to the transcriptional start site. Positions of deletions, the line representing the DNA remaining, are shown with the concomitant effect of the activity of the promoter. (Modified from An *et al.*, 1986 and Ow *et al.*, 1987.)

remaining unchanged between transformants, apparently being unaffected by position effects (Velten and Schell 1985).

Although the T-DNA genes are generally thought of as being expressed in a constitutive manner there is at least one exception to this. Gene transfer has demonstrated that the promoter of gene 5 from the T_L DNA of pTiAch5 when fused to the *ocs* gene directed its expression principally in callus, stems and petioles but not in fully developed leaves (Koncz and Schell 1986). This raises the interesting question of how the T-DNA evolved to achieve differential tissue-specific expression of its genes in plant cells and why the gene 5 promoter does not apparently function in leaf tissue.

Although the normal host range of CaMV is the genus *Brassica*, the 19S and 35S RNA promoters (Chapter 4) have both been used to direct the expression of foreign genes in a variety of plant species (Koziel *et al.*, 1984; Shewmaker *et al.*, 1985). Expression from the 35S RNA promoter appears to be higher than from the 19S RNA promoter and has been studied in most detail (for example see Ow *et al.*, 1987). Initial studies involved sub-cloning the fragment containing the 35S RNA promoter from the CaMV genome and then fusing 5'-deleted variants of it to a human growth hormone gene (*hgh*). A Ti plasmid vector was then used to insert the deleted promoter segments individually into the tobacco genome (Odell *et al.*, 1985). The levels of expression of each construct were assessed by comparing the levels of *hgh*-specific transcripts with the transcripts derived from a *nos* promoter–NPTII gene fusion following Northern blot analysis of poly(A) RNA isolated from transgenic tissue. This work demonstrated that sequences up to 46 bp upstream from the transcriptional start site were able to produce a low level of correctly initiated RNA and that the levels of transcript were increased approximately 20-fold if an upstream region of up to 168 bases was used. This led to the notion that this upstream region (-168 to -46) contained an enhancer element which served to increase the level of RNA transcribed from the promoter. The analysis of the 35S RNA promoter has been extended further by studying the transient expression in carrot cells of constructs of the deleted promoter fused to the firefly *lux* gene (Ow *et al.*, 1987). These experiments showed that the 35S RNA promoter is composed of at least three regions—distal, medial and proximal—which not only contain sequence elements which are generally considered to be important in eucaryotic gene expression but also have dramatic effects on gene expression (Fig. 6.1B). The distal region can increase the rate of transcription from a heterologous weak promoter indicating that it is indeed an enhancer element and conversely the proximal region alone can be fused to a light-induced enhancer element (see later) and direct expression in a light-dependent, tissue-specific manner (Fluhr *et al.*, 1986). Multimers of the distal region can also induce higher levels of expression than the native 35S RNA promoter. This has been confirmed in transgenic plants where strength of the 35S RNA promoter appears to be approximately 10 times higher than the *nos* promoter and when the 35S RNA promoter is present as a dimer (-343 to $-90::-343$ to $+9$) this strength is increased a further 10-fold (Kay *et al.*, 1987).

Storage Proteins

Storage proteins are synthesized in large amounts in developing seeds and tubers under precise temporal and tissue-specific control. A large number of seed proteins from a variety of plant species have been sequenced and comparisons of the regions upstream from related seed protein genes show similarities which may be important in determining the control of their transcription. Gene transfer has demonstrated that when a fragment of French bean genomic DNA containing a storage protein gene, phaseolin, flanked at the 5' and 3' ends by 863 bp and 1226 bp, respectively, was transferred into the tobacco genome, no expression of the gene could be detected in leaf tissue but phaseolin accumulated to levels greater than 1% of the embryonic protein in the tobacco seed (Sengupta-Gopalan *et al.*, 1985). This dramatic result is made all the more interesting since in bean, phaseolin accumulates in protein bodies in the cotyledons of the embryo whereas in tobacco, seed storage proteins normally accumulate in protein bodies in both the embryo and the endosperm. In the transgenic tobacco seed the phaseolin accumulated specifically in the protein bodies of the embryo. In addition, accumulation of phaseolin in the transgenic seed appears to start at about the same time in seed development as it does in the bean seed, approximately 15 days following pollination. Hence, it would seem that the controls acting on the expression of seed protein genes are highly conserved throughout evolution and that the 863 bp of sequence upstream from the phaseolin gene is sufficient to direct the correct tissue-specific, temporally-regulated expression of the gene. This work has been confirmed and extended by the transfer of the gene encoding the α-subunit of the soybean β-conglycinin seed protein into *Petunia*. In this case deletion analysis indicated that 257 bp of upstream sequence was sufficient to direct the correct expression of the gene (Beachy *et al.*, 1985).

Subsequent work, where a 17 bp fragment of soybean DNA containing four genes, including a seed lectin, was inserted into tobacco, demonstrated that differential, tissue-specific expression of the individual genes on the transferred fragment of DNA could take place indicating that not only could tobacco *trans*-acting factors recognize the soybean sequences but also that *cis*-acting sequences on the soybean DNA ensured that inappropriate expression of neighbouring genes did not take place (Okamuro *et al.*, 1986).

Similar work has been carried out with a gene encoding the most abundant protein of the potato tuber, patatin, and it has been found that sequences at the 5' end of the gene are sufficient to direct the tissue-specific expression of CAT in transgenic potato (Twell and Ooms 1987).

Proteins Involved in Photosynthesis

Photosynthesis is the process by which the plant uses light energy to produce fixed carbon within the chloroplast. A great many proteins, encoded either by the nuclear genome or the chloroplast genome, are involved in this process. Some of the principal proteins concerned with photosynthesis are extremely abundant in green tissue and this has simplified the study of their expression with two encoded by the nuclear genome being investigated in detail, the small subunit of ribulose bisphosphate carboxylase (*rbcS*) and the light-harvesting chlorophyll *a/b* binding protein (*lhcp*). These polypeptides are encoded by small multigene families in the nuclear genome and are synthesized as precursors in the cytoplasm and processed to the mature polypeptide during passage into the chloroplast. The expression of both the *rbcS* and *lhcp* genes is under complex control with light and the developmental stage of the cell exerting an influence not only on transcription but also translation and probably the turnover of the mature polypeptides. Generally, the control of the transcription of these gene families has been studied during the greening of etiolated tissue when plants grown in the dark are transferred to light. Under these conditions etioplasts are converted to active chloroplasts with a concomitant increase in the transcription of both the *rbcS* and *lhcp* genes. The induction of transcription during greening is probably controlled by several factors, one of which is the photoreceptor pigment phytochrome. Phytochrome can exist in two forms, the inactive Pr and the active Pfr. Pr is converted to Pfr on exposure to red light but this conversion can be reversed following exposure to far red light. Hence, if transcription of a gene is induced by red light and this induction can be reversed by further exposure to far red light, it is generally considered that phytochrome is the controlling factor. The expression of both *rbcS* and *lhcp* genes appears to be regulated at the level of transcription in greening tissue by phytochrome.

Several research groups have cloned the genes encoding *rbcS* and *lhcp* from a variety of species and these have been studied in detail in transgenic tissue. For example, a 280 bp upstream fragment, -330 to -50, from the pea *rbcS-3A* gene when fused to a CAT gene will direct the expression of CAT in transgenic plants in a tissue-specific manner which is regulated by phytochrome (Fluhr *et al.*, 1986). The sequence responsible for light induction has been defined further as being between -327 to -48. When this fragment is ligated to a truncated 35S RNA promoter, which is deleted at -46 and unable in itself to initiate high levels of expression (Fig. 6.1B), it can direct the tissue-specific and light-induced expression of a CAT gene located downstream (Fig. 6.2A). This induction can take place regardless of the orientation of the -327 to -48 fragment indicating that this has the properties of an enhancer element (Fluhr *et al.*, 1986).

Similar results have been obtained with the upstream element of a *lhcp* gene (Simpson *et al.*, 1986a). Here the upstream region of the pea *lhcpAB80* gene was shown to contain elements characteristic of a light-induced tissue specific enhancer between -347 and -100 with respect to the initiation of transcription. This was assessed by cloning the region upstream from a *nos* promoter–NPTII gene fusion and transferring this to the tobacco genome (Fig. 6.2B). Not only did this fragment enhance the expression of NPTII but it could also decrease, in a tissue-specific manner, the expression of the *nos* promoter. Moreover, it was demonstrated that the expression of the enhancer was dependent on the presence of functional chloroplasts in the cell (Simpson *et al.*, 1986b). In addition, it has been found that a *lhcp* gene from wheat, a monocotyledonous plant (*whAB1.6*), can function in a light-regulated and organ-specific manner in tobacco, a dicotyledonous plant (Lamppa *et al.*, 1985).

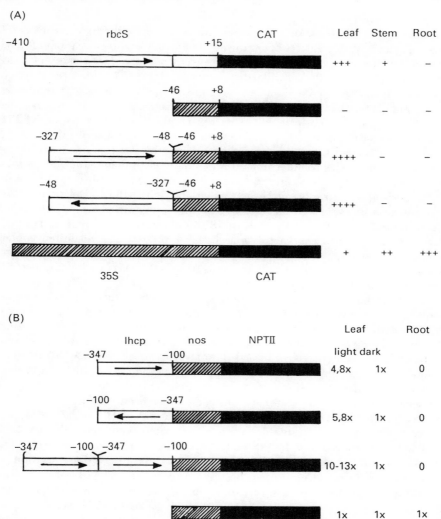

Fig. 6.2 The ability of regions upstream from genes encoding proteins involved in photosynthesis to enhance gene expression. (A). The upstream region from the pea *rbcS* gene. The levels of expression estimated by visual examination of autoradiograph of dot blot hybridized with CAT-specific hybridization probe. (B). The upstream region of the pea *lhcp* gene. Numbers refer to the position of the sequence with regard to the initiation of transcription. (Modified from Fluhr *et al.*, 1986 and Simpson *et al.*, 1986a.)

Proteins Associated with Stress

High temperature stress, or heat shock, induces the synthesis of an array of polypeptides in a wide variety of organisms. These polypeptides are thought to be involved in the acquisition of thermotolerance. Comparison of the amino acid sequences shows structural similarities between plant and animal heat shock proteins with DNA sequences in the 5′ regions of the heat shock genes showing significant conservation (Schöffl *et al.*, 1986). This is borne out by the remarkable observation that the *Drosophila* heat shock promoter, *hsp70*, fused to the NPTII gene can direct the heat induced expression of NPTII in tobacco (Spena *et al.*, 1985) and at present is the only animal promoter known to function correctly in plant cells. The *hsp70* promoter functions in a manner similar to the endogenous heat shock promoters directing expression in roots, stems and leaves but not in pollen derived from transgenic tissue (Spena and Schell 1987). Deletions of the soybean heat shock protein gene *hs6871* promoter in transgenic tobacco have shown that expression requires at least 181 bp upstream from the translational start site for correct thermal activation of gene expression and for full promoter activity 439 bp of upstream sequence are required. This upstream sequence contains similar sequence repeats to those seen in *Drosophila* (Baumann *et al.*, 1987).

Upon flooding, maize seedlings suppress the synthesis of their normal range of polypeptides and the synthesis of about 20 novel proteins is induced. Amongst these proteins are the enzymes involved in anaerobic fermentation including alcohol dehydrogenase (ADH). There are two ADH genes in maize, *Adh-1* and *Adh-2*, and their induction is controlled at the level of transcription. As described in Chapter 5 the *Adh-1* promoter fused to the CAT gene and the *nos* poly(A) addition site can direct the expression of CAT in protoplasts under low oxygen tension following electroporation. Similar constructs have been engineered into the tobacco genome by *Agrobacterium*-mediated transformation (Ellis *et al.*, 1987). In this case the construct was unable to direct the expression of detectable levels of CAT following anaerobic induction of the transgenic tissue. The precise reason for this is not clear and may be related to the efficiency of transcription initiation or the stability of the mRNA in a novel host. However, if the -395 to -86 region of the 35S RNA promoter or the upstream region of *ocs* gene was inserted upstream from the *Adh-1* promoter in the construct, high levels of CAT activity were obtained following anaerobic induction; and it was shown that a 247 bp sequence immediately upstream from the ADH gene contained all the signals necessary for anaerobic induction and accurate transcription in tobacco.

Proteins Involved with Plant–Bacterial Interaction

After considering tumour induction by *Agrobacterium*, another example of a plant/bacterial interaction is the symbiosis between legumes and *Rhizobium* spp., resulting in the formation of root nodules involved in nitrogen fixation. The bacterial–host interaction is complex with the bacteria providing many of the enzymes required for converting atmospheric nitrogen to ammonia and a number of the genes required for the symbiosis. The plant supplies both the microenvironment for nitrogen fixation and contributes the enzymes to assimilate the reduced nitrogen. The establishment of nitrogen-fixing nodules is the result of the interaction between the plant and the bacterium and involves inducing the synthesis of approximately 30 plant genes known collectively as nodulins. The induction of nodulin expression is under developmental control and although the majority have yet to be identified they include leghaemoglobin which acts as an oxygen scavenger, glutamine synthase, xanthine dehydrogenase and uricase which are involved in the metabolism of ammonium ions. Leghaemoglobin has been studied in most detail and in soybean the genes encoding it are found to be contained in two clusters in the nuclear genome. Study of the molecular basis for the induction of leghaemoglobin synthesis in particular, and the other nodulins in general, has required, in the first instance, the transformation of a legume. This has been achieved with the production of transgenic *Lotus corniculatus* using vectors based on the Ri plasmid (Chapter 3). When a construct containing the CAT gene flanked by the 5' and 3' regions of the soybean leghaemoglobin *lbc3* gene was inserted into plants, a high level of CAT activity could be detected but only in the root tissue at the precise stage of nodule development at which leghaemoglobin is normally produced (Jensen *et al.*, 1985). Subsequent work involving the comparison of CAT expression in transgenic *Lotus* containing different deletions of the *lbc3* promoter fused to the CAT gene and the *lbc3* 3' flanking sequence has shown that there are at least two regions upstream from the *lbc3* gene important in controlling the levels of expression. One element, between -230 and -170 is defined as the minimum sequence required for detectable promoter activity whereas another region between -1100 and -950 acts as an enhancer directing a 20-fold increase in the level of transcription. A 37 bp region located between -139 and -102 is sufficient to confer nodule-specific expression (Stougaard *et al.*, 1987).

Proteins Involved in Defence
It has become increasingly clear that plants synthesize a number of compounds involved in defence against both UV light and pathogens. One method by which the plant carries this out is in the induction of enzymes involved in the synthesis of flavanoids which can have an antifungal activity (phytoalexins) or serve as a UV screen to protect the plant. A key enzyme in flavanoid biosynthesis is chalcone synthase (*chs*). The *chs* gene has been isolated and characterized and a chimeric gene consisting of the *chs* promoter from *Antirrhinum majus*, the NPTII gene and the termination region of a *chs* gene from *Petroselinum hortense* (parsley) has been inserted into tobacco (Kaulen *et al.*, 1986). The *chs* promoter functioned normally in tobacco, being induced by UV-B irradiation. Deletion of the promoter sequences demonstrated the presence of two regulatory regions: one, located between -1200 and -357, was essential for optimal expression of the gene and the other, located between -357 and -39, was important in the UV-B light response.

When leaves of plants in the Solanaceae and in the Leguminosae are wounded they synthesize a protease inhibitor inducing factor (PIIF) which is transported primarily upwards from the wound site, and in turn induces the synthesis of serine protease inhibitors. These protease inhibitors are specific for non-plant proteases and are thought to interfere with the digestive systems of attacking pests. The protease inhibitor II gene has been isolated from potato and transferred to the tobacco genome. When the transferred gene is flanked at its 5' and 3' ends by 3 kb and 1.45 kb of its normal sequence respectively it could be induced not only in the wounded tissue but also throughout the plant in a similar manner to that seen in potato (Sanchez-Serrano *et al.*, 1987). Tobacco does not appear to contain sequences similar to the potato protease inhibitor II but this result suggests that upon wounding, the tobacco plant releases signals to induce the synthesis of defensive proteins and that these are sufficient to induce the transcription of the potato gene. Similar results have been obtained by fusing the potato protease inhibitor promoter to the CAT gene, transferring this to tobacco and assessing expression of the chimeric gene following wounding (Thornburg *et al.*, 1987). These workers found that the 3'-untranslated region of the gene was absolutely essential for the correct expression of the chimeric construct in the transgenic tissue, raising the point that regions downstream from the gene are also important in controlling gene expression.

Summary of Work Involving Promoter Analysis

The picture that emerges from all these studies is that the sequences upstream from plant genes are essential for their correct expression and that these regions can be incorporated into any recombinant DNA construct so that, theoretically, any chimeric gene can be expressed in a particular tissue under predetermined temporal control. At the practical level this means that any foreign gene can be inserted into plants and, as long as it is provided with the appropriate upstream DNA sequences, it is probable that it will be expressed when and where it is required. Overall, this work has pointed to the regions important in controlling gene expression and demonstrated directly that DNA sequences, which are similar in some cases to those found in other experimental systems, are indeed important in the regulation of gene expression in plants and are possibly involved in the binding of proteins which function as *trans*-acting factors to regulate when a gene is expressed and to what level it is transcribed.

Analysis of the Controls of Transcription Termination, Splicing and Sequences Important in Translation Initiation

When compared with the number of studies carried out on promoters, very little work has been reported concerning other factors important in the expression of foreign DNA in plants. These factors might include the termination of transcription, the splicing of introns and the initiation of translation. Often, chimeric genes transferred to plants contain the transcription termination signals of T-DNA genes. Construction of these regions does not appear to have to be as precise as in promoter construction and relatively large sequences of DNA can be tolerated between the translation stop site and the transcription termination signal (see for example Fig. 3.1A, p. 26). Many of the genes that have been transferred to the plant genome either do not contain introns or have been cloned as cDNAs so the question of splicing does not arise. However, there have been several examples of genes which contain introns being transferred to the tobacco genome and these include phaseolin and patatin which contain three and six introns, respectively. Processing of the RNA transcripts of these genes appears to be normal not only when transferred to a novel host (Sengupta-Gopalan *et al.*, 1985) but also when expressed in inappropriate tissue (Rosahl *et al.*, 1987). On the other hand, when a complete human growth hormone gene (*hgh*) was introduced into plant cells no splicing of the four introns was observed and the *hgh* poly(A) addition site was not used although the *nos* termination site which had been inserted downstream was (Barta *et al.*, 1986). The observation that the *hgh* pre-mRNA synthesized in plant cells could be spliced in a HeLa cell nuclear extract suggested that although the pre-mRNA was functional, the plant cells were unable to recognize the processing signals. It also appears that a dicotyledonous cell is unable to recognize the splicing signals used in monocotyledonous cells. When a wheat *rbcS* gene and a maize alcohol dehydrogenase gene were transferred to the tobacco genome, they were inefficiently spliced and although the wheat poly(A) addition site was used, other novel sites were also utilized for polyadenylation (Keith and Chua 1986).

With the aim of optimizing the expression of a foreign gene, in particular the initiation of translation of a chimeric construct, the effect of different nucleotides close to the translation initiation codon has been investigated. This was carried out using a gene fusion that consisted of a chitinase gene from *Serratia marcescens* fused to the *nos* promoter and 3′ termination sequences (Taylor *et al.*, 1987). Site-directed mutagenesis of the region surrounding the translation initiation codon was carried out and the different constructs were transferred to tobacco cells by co-cultivation and the levels of chitase-specific mRNA and protein accumulating in transgenic callus measured. It was found that changes in the nucleotides at positions -3 and $+4$ with respect to the initiation codon could lead to an 8-fold increase in the amounts of protein in the tissue and that these changes, with an additional change at $+5$, could also increase the levels of mRNA 2-fold. Hence, it appears that the nucleotides surrounding the initiation codon can affect the steady-state levels of the mRNA in transgenic cells as well as its translational efficiency.

Gene Transfer to Study Protein Function

Not only can gene transfer be used to analyse promoter sequences but where a changed phenotype is observed it can be used to identify the function of the products of a particular gene. Initially the transfer of single genes to the plant cell utilizing vectors based on the Ti plasmid was used to test the contribution of each of the *onc* genes of the Ti plasmid in tumour formation. For example, gene 4 (*ipt*) of the T-DNA introduced alone into tobacco produced tumours that spontaneously made shoots and similar work with *iaaM* and *iaaH* showed that both were sufficient to induce tumour formation on tobacco plantlets (Inze *et al.*, 1984). This type of experiment has been extended to the T-DNA of the Ri plasmid and has been used to demonstrate that *rolA*, *B* and *C* (Chapter 2) are sufficient to induce the hairy root phenotype in transformed tobacco plants (Cardarelli *et al.*, 1987; Spena *et al.*, 1987).

This type of approach has also been used to demonstrate that the tuber protein, patatin, contains an enzymatic activity. In this case it had been speculated that patatin, the major protein component of potato tubers, might contain a lipid acyl hydrolase activity. In order to confirm this a patatin gene was fused to a leaf-specific promoter, transferred into tobacco plants and the resultant leaf tissue indeed contained a unique lipid acyl hydrolase activity that co-migrated with patatin following non-denaturing polyacrylamide gel electrophoresis (Rosahl *et al.*, 1987).

In addition to investigating the function of a protein, gene transfer can also be used to analyse the functional domains of a polypeptide. Examples of this are provided by work carried out concerning the transfer of proteins to different subcellular compartments. As already described in this chapter, many proteins involved in photosynthesis are synthesized in the cytoplasm and are transported to the chloroplast. These proteins are synthesized as precursors containing an amino-terminal extension known as the transit peptide. The precursor polypeptide is translocated, post-translationally, into the chloroplast by an energy-dependent mechanism and the transit peptide removed by a protease located in the soluble fraction of the chloroplast during or shortly after the entry of the precursor into the organelle. Fusion of the transit peptide, the first codon of a pea *rbcS* gene and five artificial codons to NPTII gene followed by transfer to the tobacco genome showed that a chimaeric precursor polypeptide was synthesized and NPTII was directed into the chloroplasts in transgenic plants (Van den Broeck *et al.*, 1985). This demonstrated that the transit peptide was sufficient to target a polypeptide to the chloroplast and that processing of a chimeric precursor polypeptide could take place within the organelle. Subsequently, this work has been extended in order to study, *in vivo*, the regions of the transit peptide that are important in protein targeting and processing (Kuntz *et al.*, 1986) and it has been demonstrated that in order to obtain efficient uptake of the fusion protein into the chloroplast, the NPTII gene must be fused not only to the sequence encoding the transit peptide but also the first 23 amino acids of *rbcS* (Wasmann *et al.*, 1986).

Similar results have been obtained concerning the targeting of proteins to the mitochondria (Boutry *et al.*, 1987). In this case the transit peptide of the nuclear gene encoding the β-subunit of the mitochondrial ATPase (*atp2-1*) from *Nicotiana plumbaginifolia* was fused to a CAT gene and introduced into the tobacco genome. CAT activity was found to be preferentially associated with mitochondria isolated from transgenic plants. Although it could not be demonstrated directly that CAT was taken into the mitochondria, CAT activity was released from the organelle on sonication and processing of the precursor polypeptide by the mitochondria was demonstrated by using specific antisera raised against the protein.

Gene Transfer to Study the Molecular Biology of Viruses

The application of gene transfer to study plant viruses allows an opportunity not only to analyse viral gene expression but also the role that individual components of the viral genome may play in the viral life cycle and the production of symptoms. An example of this approach has been provided by the use of Ti plasmid mediated transformation to study the effect of Cucumber Mosaic Virus (CMV) satellite RNA on symptom expression (Baulcombe *et al.*, 1986). CMV, although biologically distinct and a member of a separate group of viruses, has a similar genomic structure and translation strategy to BMV (see Fig. 4.2, p. 54). Some isolates of the virus also contain a single-stranded linear RNA that is generally approximately 335 bases long and known as the satellite RNA. Satellite RNA does not appear to be related to CMV genomic RNA but replicates only in cells infected with CMV and when present persists as a component of the virus isolate. The presence of the satellite can modify the symptoms of infection in plants, with some isolates showing reduced symptoms whereas others have more severe symptoms. A cDNA clone representing CMV satellite RNA was inserted as a monomer or a dimer downstream from the 35S RNA promoter in a binary vector and introduced into the tobacco genome. The resultant tobacco plants did not exhibit any symptoms of viral infection and appeared normal; however, Northern blot analysis showed that they contained RNAs representing the satellite RNA sequences. The transgenic plants were subsequently inoculated with a CMV isolate lacking the satellite RNA and it was found that the inoculated virus could stimulate replication of the satellite RNA which subsequently became a component of the viral isolate. This work demonstrated for the first time that an RNA virus can acquire genetic material from its host which may be important in the evolution of plant viruses. Moreover, it was observed that the symptoms of infection in transgenic plants containing the sequences representing the satellite RNA were reduced when compared with the non-transformed plants suggesting that this might be used as a method for protecting crops against severe viral symptoms, a strategy that is discussed in more detail in Chapter 7.

Possibly the best example to date of how gene transfer can identify the function of a viral gene product has been demonstrated with the finding that the 30Kd polypeptide of TMV (see Fig. 4.1, on p. 53) potentiates cell-to-cell spread of the virus (Deom *et al.*, 1987). In order to achieve this tobacco was transformed with an SEV vector (see Chapter 3) containing a cDNA representing the 30Kd gene fused to the 35S RNA promoter. Transgenic plants transcribed the chimeric gene and the 30Kd polypeptide accumulated in the cells. When the transformed plants were inoculated with a temperature sensitive mutant of TMV which normally cannot systemically infect tobacco at non-permissive temperatures, it was found that the transgenic plants could complement the TMV mutation and systemic infection resulted. Normal tobacco plants grown and inoculated under the same conditions only produced small lesions characteristic of infection by the mutant virus. This demonstrated conclusively that the 30Kd polypeptide acts, in an as yet unknown manner to potentiate virus movement from cell to cell.

As was discussed in Chapter 4, clones representing dimers of PSTV and CaMV are infectious when inoculated onto wounded test plants. Moreover, when such dimers are inserted into T-DNA and the *Agrobacterium* containing the construct used to inoculate host plants, infection takes place. This process has been called agroinfection (Gardner *et al.*, 1986; Grimsley *et al.*, 1986a). Agroinfection is a novel method for inserting viral DNA into host plants with infection resulting either from intramolecular recombination and/or synthesis of molecules that can act as a viral replicative intermediate (Grimsley *et al.*, 1986b). It is a highly efficient and specific process requiring *vir* functions and being insensitive to DNAse present during the inoculation also suggests that it requires the normal transfer functions of the *Agrobacterium* to take place. The technique is extremely sensitive, requiring the formation of a single viral replicative intermediate in the host cell for subsequent infection to be initiated and this can take place in the absence of stable transformation. As such it has been used to investigate the transfer of DNA from *Agrobacterium* to the plant cell (Gardner *et al.*, 1986; Hille *et al.*, 1986). Agroinfection has also been used to study the effect of mutations on the viroid genome (Owens *et al.*, 1986) and has been used as a method of infecting maize with MSV (see Chapter 4) (Grimsley *et al.*, 1987). The latter case demonstrates that *Agrobacterium* can indeed transfer DNA to the cell of a graminaceous monocotyledonous plant.

Dimers of each of the components of the gemini virus, TGMV (see Chapter 4), have been inserted into the genome of *Petunia* (Rogers *et al.*, 1986). The resulting plants were normal but Southern blot analysis detected free circular single- and double-stranded viral DNAs in plants which contained the dimeric A component (or DNA 1). This indicated that component A encodes all of the functions necessary for replication and, since no symptoms were observed, suggests that component B (or DNA 2) is responsible for this during infection. Subsequent analysis of these plants revealed that the viral DNA was packaged into virions and that these produced infection when inoculated onto *Nicotiana benthamiana*, a host of TGMV, which had been engineered to contain the B component of the virus within its genome (Sunter *et al.*, 1987). This observation suggests that component B is also responsible for cell-to-cell spread of the virus. Interestingly, when plants engineered to contain dimers of A were crossed with plants containing dimers of B, one-quarter of the progeny developed symptoms of viral infection and infectious virus particles were produced. As yet, it is not clear how the dimer sequences give rise to freely-replicating viral DNAs. Due to the transcription pattern of the viral genome, complete transcription of the dimeric DNA seems unlikely so viral escape may be the result of intramolecular recombination or single-stranded DNA, which is capable of initiating infection, might be released during the replication of the genomic DNAs.

From this discussion it can be seen that the transfer of individual genes and viral genomes into novel hosts provides us with an invaluable tool with which to study gene expression and the structure/functional relationships of regions of viral genomes that are important in the life cycle of the virus and the production of symptoms in infected plants. All of these findings are of fundamental importance in understanding gene expression in plants and we shall see in Chapter 7 how this basic knowledge can be exploited in engineering plants to exhibit traits of agronomic importance.

References

An, G., Ebert, P.R., Yi, B-Y. and Choi, C-H. (1986). 'Both TATA box and upstream regions are required for the nopaline synthase promoter activity in transformed tobacco cells', *Mol. Gen. Genet.* **203**, pp. 245–250.

Barta, A., Sommergruber, K., Thompson, D., Hartmuth, K., Matzke, M.A. and Matzke, A.J.M. (1986). 'The expression of nopaline synthase–human growth hormone chimeric gene in transformed tobacco and sunflower callus tissue, *Plant Mol. Biol.* **6**, pp. 347–357.

Basiran, N., Armitage, P., Scott, R.J. and Draper, J. (1987). 'Genetic transformation of flax (*Linum usitatissimum*) by *Agrobacterium tumefaciens*: Regeneration of transformed shoots via a callus stage', *Plant Cell Reports* **6**, pp. 396–399.

Baulcombe, D.C., Saunders, G.R., Bevan, M.W., Mayo, M.A. and Harrison, B.D. (1986). 'Expression of biologically active viral satellite RNA from the nuclear genome of transformed plants', *Nature* **321**, pp. 446–449.

Baumann, G., Raschke, E., Bevan, M. and Schoffl, F. (1987). 'Functional analysis of sequences required for transcriptional activation of soybean heat shock gene in transgenic tobacco plants', *EMBO J.* **6**, pp. 1161–1166.

Beachy, R.N., Chen, Z-L., Horsch, R.B., Rogers, S.G. and Hoffman, N.J. (1985). 'Accumulation and assembly of soybean beta-conglycinin in seeds of transformed petunia plants', *EMBO J.* **4**, 3047–3053.

Boutry, M., Nagy, F., Poulsen, C., Aoyagi, K. and Chua, N-H. (1987). 'Targeting of bacterial chloramphenicol acetyltransferase to mitochondria in transgenic plants', *Nature* **328**, pp. 340–342.

Bytebier, B., Deboeck, F., De Greve, H., Van Montagu, M. and Hernalsteens, J-P. (1987). 'T-DNA organisation in tumor cultures and transgenic plants of the monocotyledon *Asparagus officinalis*', *Proc. Nat. Acad. Sci. USA* **84**, pp. 5345–5349.

Cardarelli, M., Mariotti, D., Pomoni, M., Spano, L, Capone, I. and Constantino, P. (1987). '*Agrobacterium rhizogenes* T-DNA genes capable of inducing hairy root phenotype', *Mol. Gen. Genet.* **209**, pp. 475–480.

Deak, M., Kiss, G.B., Koncz, C. and Dudits, D. (1986). 'Transformation of *Medicago* by *Agrobacterium* mediated transfer', *Plant Cell Reports* **5**, pp. 97–100.

Deom, C.M., Oliver, M.J. and Beachy, R.N. (1987). 'The 30-kilodalton gene product of tobacco mosaic virus potentiates virus movement', *Science* **237**, pp. 389–394.

Ellis, J.G., Llewellyn, D.J., Dennis, E.S. and Peacock, W.J. (1987). 'Maize *Adh-1* promoter sequences control anaerobic regulation: addition of upstream promoter elements from constitutive genes is necessary for expression in tobacco', *EMBO J.* **6**, pp. 11–16.

Everett, N.P., Robinson, K.E.P. and Mascarenhas, D. (1987). 'Genetic engineering of sunflower (*Helianthus anuus* L.)', *Bio/Technology* **5**, pp. 1201–1204.

Fillatti, J.J., Sellmer, J., McCown, B., Haissig, B. and Comai, L. (1987). '*Agrobacterium* mediated transformation and regeneration of *Populus*', *Mol. Gen. Genet.* **206**, pp. 192–199.

Fluhr, R., Kuhlemeyer, C., Nagy, F. and Chua, N-H. (1986). 'Organ-specific and light induced expression of plant genes', *Science* **232**, pp. 1106–1112.

Gardner, R.C., Chonoles, K.R. and Owens, R.A. (1986). 'Potato spindle tuber viroid infections mediated by the Ti plasmid of *Agrobacterium tumefaciens*', *Plant. Mol. Biol.* **6**, pp. 221–228.

Gardner, R.C. and Knauf, V.C. (1986). 'Transfer of *Agrobacterium* DNA to plants requires a T-DNA border but not the *virE* locus', *Science* **231**, pp. 725–727.

Grimsley, N., Hohn, B., Hohn, T. and Walden, R. (1986a). 'Agroinfection—an alternative route for viral infection of plants by using the Ti plasmid', *Proc. Nat. Acad. Sci. USA* **83**, pp. 3282–3286.

Grimsley, N., Hohn, T. and Hohn, B. (1986b). 'Recombination in a plant virus: template switching in cauliflower mosaic virus', *EMBO J.* 5, pp. 641–646.

Grimsley, N., Hohn, T., Davies, J.W. and Hohn, B. (1987). '*Agrobacterium*-mediated delivery of infectious maize streak virus into maize plants', *Nature* 325, pp. 177–179.

Guerche, P., Jouanin, L., Tepfer, D. and Pelletier, G. (1987). 'Genetic transformation of oilseed rape (*Brassica napus*) by the Ri T-DNA of *Agrobacterium rhizogenes* and analysis of inheritance of the transformed phenotype', *Mol. Gen. Genet.* 206, pp. 382–386.

Hille, J., Dekker, M., Oude Luttighuis, H., Van Kammen, A. and Zabel, P. (1986). 'Detection of T-DNA transfer to plant cells by *A. tumefaciens* virulence mutants using agroinfection', *Mol. Gen. Genet.* 205, pp. 411–416.

Horsch, R.B., Fraley, R.T., Rogers, S.G., Sanders, P.R., Lloyd, A. and Hoffman, N. (1984). 'Inheritance of functional foreign genes in plants', *Science* 223, pp. 496–498.

Horsch, R.B., Fry, J.E., Hoffman, N.L., Eichholtz, D., Rogers, S.G. and Fraley, R.T. (1985). A simple and general method for transferring genes into plants. *Science* 227, pp. 1229–1231

Inze, D., Follin, A., Van Lijsehettens, L., Simoens, G., Genetello, C., Van Montagu, M. and Schell, J. (1984). 'Genetic analysis of the individual T-DNA genes of *Agrobacterium tumefaciens*: further evidence that two genes are involved in indole-3-acetic acid synthesis', *Mol. Gen. Genet.* 194, pp. 265–274.

Jensen, J.S., Marcker, K.A., Otten, L. and Schell, J. (1985). 'Nodule specific expression of a chimeric soybean leghemoglobin gene in transgenic *Lotus corniculatus*', *Nature* 321, pp. 669–674.

Jones, J.D.G., Dunsmuir, P. and Bedbrook, J. (1985). 'High level expression of introduced chimeric genes in regenerated transformed plants', *EMBO J.* 4, pp. 2411–2418.

Kaulen, H., Schell, J. and Kreuzaler, F. (1986). 'Light-induced expression of the chimeric chalcone synthase–NPT-II gene in tobacco cells', *EMBO J.* 5, pp. 1–8.

Kay, R., Chan, A., Daly, M. and McPherson, J. (1987). 'Duplication of CaMV 35S promoter sequences creates a strong enhancer for plant genes', *Science* 236, pp. 1299–1302.

Keith, B. and Chua, N-H. (1986). 'Monocot and dicot pre-mRNAs are processed with different efficiencies in transgenic tobacco', *EMBO J.* 5, pp. 2419–2424.

Koncz, C. and Schell, J. (1986). 'The promoter of the TL-DNA gene 5 controls the tissue specific expression of chimeric genes carried by a novel type of *Agrobacterium* binary vector', *Mol. Gen. Genet.* 204, pp. 383–396.

Koziel, M.G., Adams, T.L., Hazlet, M.A., Damm, D., Miller, J., Dahlbeck, D., Jayne, S. and Staskawitcz, B. (1984). 'A cauliflower mosaic virus promoter directs the expression of kanamycin resistance in morphogenic transformed plant cells', *J. Mol. Appl. Genet.* 2, pp. 549–562.

Kuntz, M., Simons, A., Schell, J. and Schreier, P. (1986). 'Targeting of protein to chloroplasts in transgenic tissue by fusion with mutated transit peptide', *Mol. Gen. Genet.* 205, pp. 454–460.

Lamppa, G., Nagy, F. and Chua, N-H. (1985). 'Light-regulated and organ-specific expression of a wheat Cab gene in transgenic tobacco', *Nature* 316, pp. 750–752.

Lloyd, A., Barnason, A., Rogers, S.G., Byrne, M., Fraley, R.T. and Horsch, R.B. (1986). 'Transformation of *Arabidopsis thaliana* with *Agrobacterium tumefaciens*', *Science* 234, pp. 464–466.

Manners, J.M. (1987). 'Transformation of *Stylosanthes* spp. using *Agrobacterium tumefaciens*', *Plant Cell Reports* 6, pp. 204–207.

McCormack, S., Niedermeyer, J., Fry, J., Barnason, A., Horsch, R.B. and Fraley, R.T. (1986). 'Leaf disc transformation of cultivated tomato (*L. esculentum*) using *Agrobacterium tumefaciens*', *Plant Cell Reports* 5, pp. 81–84.

Michelmore, R., Marsh, E., Seely, S. and Landry, B. (1987). 'Transformation of lettuce (*Lactuca sativa*) mediated by *Agrobacterium tumefaciens*', *Plant Cell Reports* **6**, pp. 439–442.

Noda, T., Tanaka, N., Mano, Y., Nabeshima, S., Ohkaawa, H. and Matsui, C. (1987). 'Regeneration of horseradish hairy roots incited by *Agrobacterium rhizogenes* infection', *Plant Cell Reports* **6**, pp. 283–286.

Odell, J.T., Nagy, F. and Chua, N-H. (1985). 'Identification of DNA sequences required for the activity of the cauliflower mosaic virus 35S promoter', *Nature* **313**, pp. 810–812.

Okamuro, J.K., Jofuka, K.D. and Goldberg, R.B. (1986). 'The soybean lectin gene and flanking nonseed protein genes are developmentally regulated in transformed tobacco plants', *Proc. Nat. Acad. Sci. USA* **83**, pp. 8240–8244.

Ooms, G.G., Burrell, M.M., Karp, A., Bevan, M. and Hille, J. (1987). 'Genetic transformation in two potato cultivars with T-DNA from disarmed *Agrobacterium*', *Theor. Appl. Genet.* **73**, pp. 744–750.

Ow, D.W., Jacobs, J.D. and Howell, S.H. (1987). 'Functional regions of the Cauliflower Mosaic Virus 35S RNA promoter determined by the use of the firefly luciferase gene as a reporter of promoter activity', *Proc. Nat. Acad. Sci. USA* **84**, pp. 4870–4874.

Owens, R.A., Hammond, R.W., Gardner, R.C., Kiefer, M.C., Thompson, S.M. and Cress, D.E. (1986). 'Site specific mutagenesis of potato spindle tuber viroid', *Plant Mol. Biol.* **6**, pp. 179–192.

Petit, A., Stougaard, J., Kuhle, A., Marcker, K.A. and Tempe, J. (1987). 'Transformation and regeneration of the legume *Lotus corniculatus*: A system for molecular studies of symbiotic nitrogen fixation', *Mol. Gen. Genet.* **207**, pp. 245–250.

Rogers, S.G., Bisaro, D.M., Horsch, R.B., Fraley, R.T., Hoffmann, N.L., Brand, L., Elmer, J.S. and Lloyd, A.M. (1986). 'Tomato golden mosaic virus A component DNA replicates autonomously in transgenic plants', *Cell* **45**, pp. 593–600.

Rosahl, S., Schell, J. and Willmitzer, L. (1987). 'Expression of a tuber-specific storage protein in transgenic tobacco plants: demonstration of an esterase activity'. *EMBO J.* **6**, pp. 1155–1159.

Sanchez-Serrano, J.J., Keil, M., O'Connor, A., Schell, J. and Willmitzer, L. (1987). 'Wound induced expression of a potato proteinase inhibitor II gene in transgenic tomato plants', *EMBO J.* **6**, pp. 303–306.

Sengupta-Gopalan, C., Reichert, N.A., Barker, R.F., Hall, T.C. and Kemp, J.D. (1985). 'Developmentally regulated expression of the bean β-phaseolin gene in tobacco seed', *Proc. Nat. Acad. Sci. USA* **82**, pp. 3320–3324.

Schafer, W., Gorz, A. and Kahl, G. (1987). 'T-DNA integration and expression in a monocot crop plant after induction of *Agrobacterium*', *Nature* **327**, pp. 529–532.

Schöffl, F., Baumann, G., Raschke, E. and Bevan, M. (1986). 'Expression of heat shock genes in higher plants', *Phil. Trans. R. Soc. Lond.* (*B*) **314**, pp. 453–468.

Shewmaker, C.K., Caton, J.R., Heuck, C.M. and Gardner, R.C. (1985). 'Transcription of Cauliflower Mosaic Virus integrated into plant genomes', *Virology* **140**, pp. 281–288.

Simpson, J., Schell, J., Van Montagu, M. and Herrera-Estrella, L. (1986a). 'Light inducible and tissue specific pea *1hcp* gene expression involves upstream element combining enhancer- and silencer-like properties. *Nature* **323**, pp. 551–554.

Simpson, J., Van Montagu, M. and Herrera-Estrella, L. (1986b). 'Photosynthesis-associated gene families: Difference in response to tissue-specific and environmental factors', *Science* **233**, pp. 34–38.

Spena, A. and Schell, J. (1987). 'The expression of a heat-inducible chimeric gene in transgenic tobacco plants', *Mol. Gen. Genet.* **206**, pp. 436–440.

Spena, A., Hain, R., Ziervogel, V., Saedler, H. and Schell, J. (1985). 'Construction of a heat inducible gene for plants. Demonstration of heat inducible activity of the *Drosophila hsp70* promoter in plants', *EMBO J.* **4**, pp. 2739–2742.

Spena, A., Schmulling, T., Koncz, C. and Schell, J. (1987). 'Independent and synergistic activity of *rolA*, *B* and *C* loci in stimulating abnormal growth in plants', *EMBO J.* **6**, pp. 3891–3899.

Stougaard, J., Sandal, N.N., Gron, A., Kuhle, A. and Marcker, K.A. (1987). '5' Analysis of soybean leghaemoglobin *lbc3* gene: regulatory elements required for promoter activity and organ specificity', *EMBO J.* **6**, pp. 3565–3569.

Sunter, G., Gardiner, W.E., Rushing, A.E., Rogers, S.G. and Bisaro, D.M. (1987). 'Independent encapsidation of tomato golden mosaic virus A component DNA in transgenic plants', *Plant Mol. Biol.* **8**, pp. 477–484.

Taylor, J.L., Jones, J.D.G., Sandler, S., Mueller, G.M., Bedbrook, J. and Dunsmuir, P. (1987). 'Optimizing the expression of chimeric genes in plant cells', *Mol. Gen. Genet.* **210**, pp. 572–577.

Tepfer, D. (1984). 'Transformation of several species of higher plants by *Agrobacterium rhizogenes*: Sexual transmission of the transformed genotype and phenotype', *Cell* **37**, pp. 959–967.

Thornburg, R.W., An, G., Cleveland, T.E., Johnson, R. and Ryan, C.A. (1987). 'Wound inducible expression of a potato protease inhibitor II chloramphenicol acetyltransferase gene fusion in transgenic tobacco plants', *Proc. Nat. Acad. Sci. USA* **84**, pp. 744–748.

Trulson, A.J., Simpson, R.B. and Shahin, E.A. (1986). 'Transformation of cucumber (*Cucumis sativus* L.) plants with *Agrobacterium rhizogenes*', *Theor. Appl. Genet.* **73**, pp. 11–15.

Twell, D. and Ooms, G. (1987). 'The 5' flanking DNA of the patatin gene directs tuber-specific expression of a chimeric gene', *Plant Mol. Biol.* **9**, pp. 345–375.

Umbeck, P., Johnson, G., Barton, K. and Swain, W. (1987). 'Genetically transformed cotton (*Gossypium hirsutum* L.)', *Bio/Technology* **5**, pp. 263–266.

Van den Broeck, G., Timko, M.P., Kausch, A.P., Cashmore, A.R., Van Montagu, M. and Herrera-Estrella, L. (1985). 'Targeting of a foreign protein to chloroplasts by fusion to the transit peptide from small subunit of ribulose 1,5-bisphosphate carboxylase', *Nature* **313**, pp. 358–363.

Velten, J. and Schell, J. (1985). Selection-expression plasmid vectors for use in genetic transformation of higher plants. *Nuc. Acids Res.* **13**, pp. 6981–6998.

Velten, J., Velten, L., Hain, R. and Schell, J. (1984). 'Isolation of a dual plant promoter fragment from the Ti plasmid of *Agrobacterium tumefaciens*', *EMBO J.* **3**, 2723–2730.

Wasmann, C.A., Reiss, B., Bartlett, S. and Bohnert, H. (1986). 'The importance of the transit peptide and the transported protein for protein import into chloroplasts', *Mol. Gen. Genet.* **205**, pp. 446–453.

Wei, Z-M., Kamada, H. and Harada, H. (1986). 'Transformation of *Solanum nigrum* L. protoplasts by *Agrobacterium rhizogenes*', *Plant Cell Reports* **5**, pp. 93–96.

White, D.W.R. and Greenwood, D. (1987). 'Transformation of the forage legume *Trifolium ripens* L. using binary *Agrobacterium* vectors', *Plant Mol. Biol.* **5**, pp. 461–469.

Zambryski, P., Joos, H., Genetello, C., Leemans, J., Van Montagu, M. and Schell, J. (1983). Ti plasmid vector for the introduction of DNA into plant cells without alteration of their normal regeneration capacity. *EMBO J.* **2**, pp. 2143–2150.

Chapter 7

Engineering Plants to Contain Useful Agronomic Traits

Potential Applications of Gene Transfer in Plants

The ability to transfer sequences of DNA to the plant cell, have them integrated stably into the genome and expressed, raises a wide variety of potential practical opportunities. For example, for several years there has been interest in using plant cell cultures in the production of pharmaceuticals or fine chemicals and recently attention has turned to using root cultures, induced by *Agrobacterium rhizogenes*, to produce root-derived secondary metabolites such as the anti-inflammatory drug, shikonin (for review see Hamill *et al.*, 1987). Gene transfer can also be exploited directly by engineering cell lines to produce polypeptides of pharmaceutical value. This becomes particularly attractive if the plant cell is able to carry out post-translational modifications to the protein, for example glycosylation, which other systems such as bacteria, yeast or cultured animal cells might not be able to do correctly.

Generally however, as the techniques of gene transfer have developed, most attention has been focused on their potential application in crop improvement with the aim of engineering specific traits into a wide variety of plants. Goals that are most often cited in this context include changes in the oil content or quality of seeds, the manipulation of seed proteins so that they contain a more balanced mixture of amino acids and introducing tolerance to environmental stress, pathogens or pests and herbicides. With our increasing knowledge of the genes responsible for these traits and how to arrange for the expression of a foreign gene in the correct organ at the appropriate development stage these ideas are becoming feasible. However, unfortunately gene transfer technology has advanced at a far faster pace than our understanding of plant biochemistry and of the factors which are important within the plant in determining what might be considered as other useful agronomic traits. Indeed it is difficult to conceive how gene transfer at the moment can be of aid, for example, to the cereal breeder interested in such characteristics as enhanced vigour, increased yield or resistance to lodging. Many such characteristics are likely to be determined by the interaction of many diverse gene products which are probably involved in a variety of biochemical processes and these need first to be identified before molecular biologists can attempt to isolate them. It is this area of plant biochemistry where the gap in our knowledge is the widest. The greatest difficulty is likely to be in the method of isolating the genes which individually do not confer a detectable phenotype but which are important in determining useful traits when expressed in concert with others. Because of this, attention has been focused largely on characters which might be determined by single genes and tolerance to herbicides, pathogens or pests immediately come to mind. Resistance to pathogens or pests are characters whose usefulness is self-evident and there are a variety of good reasons why herbicide tolerance might be of benefit to farmers as well as the companies who make and market the herbicides. For example, if a broad-spectrum herbicide kills all plants it can only be applied in a limited manner, prior to the germination of the crop plant. If the crop plant in question can tolerate the presence of a herbicide this increases the period in which the herbicide can be used allowing it to be applied when it is likely to be of maximum benefit, when the young seedling is competing with weeds for nutrients and radiant energy. Another potential use comes from the fact that whereas some crop plants are resistant to a herbicide, others may not be so that if the residue of the herbicide remains in the soil crop rotation may be precluded. An example of this is the rotation of maize with soybean and the use of atrazine to which the former is resistant and the latter sensitive.

Engineering Herbicide Tolerance

Herbicides can act in a variety of ways to kill plants, affecting either the chloroplast, mitochondria, nucleic acid metabolism, protein synthesis or membrane interactions (for review see Moreland 1980). In order to engineer tolerance to a herbicide into a plant, the molecular basis of its mode of action needs to be well characterized and generally this has been well researched by the agro-chemical companies. Some of the most successful herbicides act at a single step in a biochemical pathway affecting an enzymatic reaction that plays a key role in the biochemistry of the cell. For example, the site of action of chlorsulfuron and sulfometuron methyl, the respective active constituents of Glean[R] and Oust[R] produced by DuPont, is acetolactate synthase, the first enzyme specific to the synthetic pathway of the branched chain amino acids isoleucine, leucine and valine (Chaleff and Mauvais 1984).

Herbicide tolerance can be introduced into a crop plant by breeding and an example of this has been provided by the production of atrazine-resistant oilseed rape, also known as canola (*Brassica napus*). Atrazine interferes with electron transport in the chloroplast by binding to a membrane polypeptide (*psbA*) that is encoded by the chloroplast genome. Spontaneous mutants resistant to atrazine have been found in several plant species and these have minor changes in the sequence of *psbA*. The herbicide-resistant canola was produced by breeding oilseed rape with an atrazine-resistant relative, birds rape (*Brassica campestris*) (Beversdorf *et al.*, 1980). This approach, however, is not feasible with most crop plants since they do not have the natural variability for herbicide tolerance and this has prompted interest in engineering resistance. Atrazine is known to be detoxified by glutathione-S-transferase, and the gene for this has been isolated from maize (Shah *et al.*, 1986a) and, potentially it could be used to confer resistance to atrazine-sensitive plants.

Generally, where the herbicide is known to act at a single biochemical step several strategies can be adopted in attempting to engineer tolerance. These are discussed below.

Strategies to Engineer Herbicide Tolerance

Selection for Resistance in Tissue Culture.   This procedure involves culturing cells in the presence of the herbicide in order to select for growth of any that are resistant. Thus it is possible to screen large numbers of cells for variants which can be induced either by somaclonal variation or by mutagenesis. Once putative resistant cells have been selected they can be regenerated into plants with the hope that the regenerated plant will retain the resistant phenotype.

Over-production of the Target Polypeptide.   Should a single polypeptide be the target of the herbicide the effects of the herbicide might be overcome if the protein is synthesized at enhanced levels, thereby simply swamping the herbicide in the plant cell. This approach is obviously most feasible if the herbicide is present at low levels in the cell and binds irreversibly to the protein although this is not an absolute necessity.

Isolating Genes Encoding Polypeptides Resistant to the Herbicide.   If a gene encoding a target polypeptide which is resistant to the herbicide but retains its enzyme activity can be isolated, this can be used to confer tolerance following transfer to a sensitive host. One example of this is provided by a mutant tobacco acetolactate synthase gene which has been used to confer resistance to sulfonylurea herbicides (Mazur *et al.*, 1987). Another example is provided by the case of the atrazine-resistant *psbA*, where the gene might be isolated from herbicide resistant plants and used to confer tolerance in a normally sensitive host. Some herbicides act at steps in biochemical pathways that are so fundamental to living organisms that the enzyme has been conservd throughout evolution and may be present in bacteria. Should this be the case, then the enzyme from a bacterium may be resistant to the herbicide.

Engineering a Resistant Polypeptide.   If resistance to a herbicide can be screened effectively in bacteria and if the structure of the target peptide as well as the mode of action of the herbicide are known, the protein can be modified by *in vitro* mutagenesis using standard recombinant DNA techniques to produce a herbicide tolerant enzyme. Theoretically, this method allows the modification of discrete portions of the polypeptide which may affect the site of action of the herbicide without changing the active site of the enzyme itself if these two regions are separate.

Detoxifying Enzymes.   Rather than attempting to change the target site of the herbicide this approach aims at modifying the herbicide within the cell before it can act on its target. In order to do this an enzyme capable of detoxifying the herbicide needs to be isolated. This is most easily approached by screening plants that are resistant to the herbicide or the microorganisms that are found in herbicide-contaminated soils and which degrade the active ingredient of the herbicide. In this approach care needs to be taken to demonstrate that tolerance results from detoxification and not the presence of resistant enzymes. The approach is simplified if there is a single detoxifying enzyme with a high affinity to the herbicide. It is also possible that the active ingredient of the herbicide is synthesized by a microorganism. In such circumstances it is likely that the same microorganism may also produce a detoxifying enzyme to protect itself against the toxic affect of the compound. With this approach it is important that the degradation products of the herbicide are not toxic to humans or grazing animals.

Of the work aimed at engineering herbicide resistance into plants using recombinant DNA technology and gene transfer, two projects have been reported in detail in the scientific press: engineering resistance to glyphosate and to phosphinothricin. These are discussed below.

Glyphosate Tolerance
Glyphosate, the active ingredient of the herbicides Roundup^R and Tumbleweed^R, is a broad spectrum, non-selective herbicide effective against many plants. It is rapidly absorbed by the plant and is considered to be environmentally acceptable as it is non-toxic to animals and is rapidly degraded by soil microorganisms. The target for the herbicide is 5-enol-pyruvylshikimate-3-phosphate (EPSP) synthase, a nuclear encoded enzyme central to the shikimate pathway involved with aromatic amino acid synthesis. EPSP synthase activity is located primarily within the chloroplasts. Initially attempts to engineer resistance involved cell culture and it was found that a glyphosate tolerant cell line of *Petunia hybrida* contained amplified copies of the EPSP synthase gene which resulted in its over-expression. This not only suggested that over-expression of EPSP synthase might prove to be a useful strategy in engineering tolerance but it also eased the synthesis and isolation

Fig. 7.1 Glyphosate tolerant transgenic plants. *Petunia* engineered to over-express EPSP synthase were sprayed with Roundup[R] at a dose equivalent of 0.9 kg/ha. Transgenic plants (upper) and control plants (lower) were photographed three weeks after spraying. (Reproduced with permission from Shah *et al.*, 1986b).

of a cDNA representing EPSP synthase. As described in Chapter 3 this cDNA has been used to engineer tolerance to glyphosate at levels of 0.9 kg/ha in *Petunia* by fusing the gene to the 35S RNA promoter (see Fig. 7.1). The cDNA encoded not only the mature protein but also the transit peptide so that the polypeptide could be taken up by chloroplasts *in vitro* and processed to produce the mature peptide (della-Chioppa *et al.*, 1986).

Another approach has been to isolate a mutant gene encoding EPSP synthase (*aroA*) from *Salmonella typhimurium* which is resistant to glyphosate and introducing this into plants (Comai *et al.*, 1985). *AroA* was fused to either the *ocs* or the mannopine synthase promoter and introduced into tobacco plants. Because *aroA* is not a secretory protein in *Salmonella* it was thought that the EPSP synthase was located in the cytoplasm of the transgenic plants, nevertheless they displayed tolerance to glyphosate with the levels of resistance being correlated with the levels of expression of the *aroA* gene. The *aroA* construct has also been introduced into poplar (Fillatti *et al.*, 1987b) and tomato (Fillatti *et al.*, 1987a). In the latter case, plants resistant to a level of 0.84 kg/ha glyphosate were obtained.

It is thought that tolerance to higher levels of glyphosate can be achieved if the mutant EPSP synthase enzymes from microorganisms can be targeted specifically to the chloroplast. In order to achieve this the glyphosate resistant *aroA* gene has

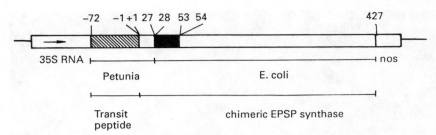

Fig. 7.2 Targeting a mutant *aroA* gene to the chloroplast. The chimeric
gene is flanked by the 35S RNA promoter and the *nos* poly(A) addition site.
A portion of the glyphosate resistant *aroA* gene from *E. coli* is fused to a
portion of the wild type *aroA* gene, the region encoding the first 27 amino
acids of the *Petunia* EPSP synthase and the transit peptide. The filled block
represents the portion of wild type *E. coli aroA*, the numbers refer to the
position of the amino acids in the chimeric construct. (Modified from della-
Chioppa *et al.*, 1987).

also been isolated from *E. coli* and fusions made between a portion of the wild type
E. coli gene, the region encoding the first 27 amino acids of the *Petunia* EPSP
synthase and the transit peptide, thus optimizing transfer of the enzyme to the
chloroplast (see Fig. 7.2). The chimeric enzyme produced by this construct was
found to be taken up and processed by chloroplasts both *in vitro* and *in vivo* (della
Chioppa *et al.*, 1987). This is a further example of a foreign protein being targeted
to the chloroplast and in this case the chimeric protein can form a conformation
which is enzymatically active.

Tolerance to Phosphinothricin
Phosphinothricin (PPT), an analogue of glutamine, is the active ingredient of two
potent herbicides, Basta[R] and Herbiace [R]. The former is made by the chemical
synthesis of PPT whereas the latter is produced by the fermentation of *Streptomyces
hygroscopicus* which produces 'bialaphos', a tripeptide of PPT and two L-alanine
residues. Within both bacteria and plants intracellular peptidases remove the
alanine residues to release PPT. PPT acts as an inhibitor of glutamine synthase
(GS) which in higher plants is important in the assimilation of ammonia.
Analogous work to that previously described with glyphosate, involving the
selection of resistant cells from tissue culture, showed that a PPT-tolerant alfalfa
cell line contained amplified copies of the GS gene and increased levels of GS
activity in the cells. Thus it is possible that over-expression of GS in plant cells
might result in enhanced tolerance to PPT.

Another approach has been taken by engineering plants to synthesize a protein
which can detoxify PPT (DeBlock *et al.*, 1987). In this case advantage was taken of
the observation that *S. hygroscopicus*, which synthesizes bialaphos, contains a
phosphinothricin acetyltransferase (PAT) activity which acetylates the free
amino group of PPT thus detoxifying it. The bialaphos resistance gene (*bar*) has
been isolated and appears to be involved in both the biosynthesis and the

prevention of accumulation of PPT (Thompson *et al.*, 1987). The *bar* gene has been engineered to be expressed in plant cells by replacing the translation initiation codon used in *Streptomyces*, GTG, by ATG and fusing it to the 35S RNA promoter and the octopine T_L gene 7 termination sequence. As described in Chapter 3, PAT can be used as a selectable marker and the transgenic tobacco plants obtained are resistant to both Basta[R] and Herbiace[R]. The resistance has also been transferred to potato and tomato with the production of Basta[R]-tolerant transgenic plants.

Engineering Resistance to Insects

Two approaches have been used to engineer resistance to insects into plants. One capitalizes on the observation that the bacterium *Bacillus thuringiensis* synthesizes a polypeptide (B.t. toxin) which is lethal to insects and the other utilizes the natural defence mechanisms of plants against insects.

B.t. toxin resides as a crystal in a parasporal inclusion body which forms during sporulation of the bacterium. Different strains of the bacterium synthesize toxins with different specificities. For example, varieties *berliner* and *kurstaki* produce toxins which are active against Lepidoptera (e.g. moths) whereas variety *isrealensis* toxin is active against Diptera (e.g. mosquitoes and blackfly). The toxins, ingested by the target insect are solubilized and cleaved by proteases active in the gut to produce the active toxin. An early sign of intoxication is a loss of appetite by the larvae. This is followed by death, although the exact cause of death is not known. Spore preparations containing a mixture of spores and the toxin from *B. thuringiensis* have been marketed for some time as 'natural pesticides' although the use of such preparations can be expensive and the protection against insects is dependent on the stability of the toxin after field application. A method to obviate these difficulties would be to engineer a plant to synthesize its own toxin.

In order to investigate this possibility the gene from *B. thuringiensis* var. *berliner* encoding a 1155 amino acid protein has been isolated and characterized. Even though this protein is a pretoxin it is toxic to lepidopteran larvae and when it is cleaved to a smaller polypeptide of 60 Kd it retains full activity. Deletions of the gene have been constructed *in vitro* and used to synthesize truncated proteins in *E. coli*. Thus it could be shown that the portion of the polypeptide which is essential for toxicity is located in the amino terminal region of the protein between amino acids 29–607. Bearing this information in mind, both the gene encoding the normal polypeptide and a truncated gene encoding the active portion of the protein were cloned into an expression vector based on the dual promoter of genes 1 and 2 of the T_R DNA of pTiAch5 (see Chapter 6). The promoter sequence of gene 2 directed the expression of the toxin and the promoter of gene 1 directed the expression of the NPTII gene. In addition, translational fusions were made between the active regions of the toxin and the NPTII gene so as to produce a chimeric polypeptide that would not only confer resistance to kanamycin but also be toxic to lepidopteran larvae (see Fig. 7.3). These constructs were introduced into tobacco and transformants were selected by the resistance to high

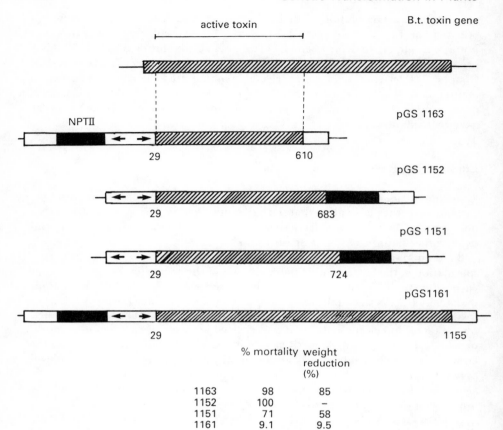

Fig. 7.3 Engineering the B.t. toxin gene for expression in plants. The
structure of the expression cassette for the expression of B.t. toxin in plant
cells. The divergent arrows represent the T_R dual promoter, the filled box
represents the NPTII gene; the B.t. toxin gene is represented by the hatched
area; the open boxes refer to the poly (A) addition sites used, either isolated
from the *ocs* gene (left) or T_L gene 7 (right). % mortality and weight loss of
surviving larvae are averaged from the test plants used (modified from Vaeck
et al., 1987).

levels of kanamycin, the rationale being that resistance to high levels of kanamycin
might be correlated with high levels of toxin synthesis. The leaves of the transgenic
plants were fed to larvae of *Manducta sexta* (tobacco hornworm), a pest of tobacco;
mortality or weight change of the larvae were monitored. In order to do this, leaf
disks were cut, placed on filter paper and infected with the larvae in two batches.
Interestingly, no insecticidal activity could be obtained in plants containing the
full B.t. toxin gene; however, in plants containing the B.t. toxin: NPTII fusion
there was insecticidal activity which could be correlated to resistance to

kanamycin. Indeed, 75% of plants that were resistant to high levels of kanamycin ($1000 \mu g/ml$) caused 75–100% insect mortality. It is not clear why the full length B.t. toxin clone did not produce insect resistant plants, although it could be that the full length mRNA or protein is unstable in plants.

Similar results have been obtained by inserting the toxin of *B. thuringiensis* var. *kurstaki* into tomato (Fischhoff *et al.*, 1987). In this example both the complete toxin gene and a truncated variant encoding amino acids 1–725 were inserted downstream from the 35S RNA promoter and inserted into the tomato genome. Transgenic plants contained B.t. toxin-specific mRNA and were resistant to three lepidopteran pests of tomato—*Manduca sexta* (tobacco hornworm), *Heliothis zea* (corn earworm) and *Heliothis virescens* (tobacco budworm).

These results demonstrate the feasibility of using toxin genes from a bacterium to control insect predators. It is likely that examples of this approach will increase to include toxins which are known to be active against pests of major agricultural importance such as Colorado beetle (*Leptinotarsa decemlineata*) and boll weevil (*Anthonomus grandis*). This type of insect control is considered biologically acceptable because the toxins are highly specific with individual toxins being active against a single order of insects. It remains to be seen whether extensive use of this approach will lead to an establishment of a field resistance to a particular toxin by the target insect. Nevertheless the cloning of the genes will allow further studies on not only the mode of action of the toxin but also protein engineering to modify the toxins in an attempt to increase the activity of the toxin against insects that are currently insensitive to it.

This type of experiment raises the issue of how acceptable plants synthesizing an endotoxin will be commercially. For example, in tomato plants engineered to contain the B.t. toxin, low levels of endotoxin were reported to be present in the fruit which can also be susceptible to larval attack. This is perhaps to be expected considering that the toxin genes had been fused to constitutive promoters which are likely to direct high levels of expression. Because the toxin is specific to insects it is considered harmless to man and it has even been suggested that over-production of the toxin might add nutritional value to the plant as far as human nutrition is concerned (Dean 1984). Nevertheless, it may be more efficacious for the toxin genes to be linked to a wound-induced promoter so that expression is triggered specifically when insect larvae begin to feed on the plant. However, for this to be successful the induction will need to be rapid and produce sufficient levels of toxin to be lethal to the larvae.

The alternative approach is to enhance the expression of plant genes which are thought to be a defence response to insect attack and transfer these genes to plants where they are not normally present. For example, as was seen in Chapter 6, wounded plants produce a proteinase inhibitor inducing factor (PIIF) which in turn induces the synthesis of protease inhibitors which are specific against insect and microbial proteases and are thought to be important in the defence of the plant to pathogens and pests. Another group of protease inhibitors that have been well characterized are those that are present in seeds and are thought to be important in providing field resistance against insects. The cowpea trypsin inhibitor has been cloned and inserted into tobacco where it was found to result in enhancing the resistance of the plants to predation by *Heliothis virescens* (Hilder *et al.*, 1987). An advantage of using this type of strategy is that the protease inhibitor is effective against a wide range of insects some of which are of major economic importance and it appears to have no toxic effect on humans.

Engineering Symptom Reduction During Viral Infection

Two strategies have been used in engineering plants so that the symptoms of viral infection are reduced. One is an adaptation of viral cross protection whereas the other utilizes symptom modification by viral satellite RNA.

The phenomenon of viral cross protection has been known for some time and the term was first coined by McKinney in 1929 after observing that a strain of TMV that induced green mosaic symptoms could prevent another strain of TMV producing the normal yellow mosaic symptoms. Normally cross protection is considered as the ability of a virus to suppress or delay the symptoms of infection caused by a superinfecting (or challenge) virus. Cross protection has been used in crop protection (see review by Fulton 1986). However, there are difficulties with this practice, not least because of the need to inoculate a crop with a virus.

This could be obviated by engineering a plant to contain and express a viral genome or the portion of it that is responsible for cross protection. In order to investigate this a cDNA representing the coat protein gene of the U_1 strain of TMV was cloned between the 35S promoter and the *nos* terminator and inserted into the tobacco genome (Powell Abel *et al.*, 1986). The transgenic plants contained RNA representing the transcript of the coat protein and the protein itself accumulated to levels of up to 0.1% of the total protein of the leaf. When transgenic plants expressing the coat protein were inoculated with the U_1 strain of TMV, symptoms of infection did not appear or appeared more slowly when compared to inoculated control plants. Increasing the amount of TMV in the challenge inoculation decreases the effect of cross protection resulting in an increase in the number of plants infected, a phenomenon which is also seen in conventional cross protection. Subsequent work showed that plants containing the U_1 coat protein gene were also protected against infection by a more severe strain of TMV (Nelson *et al.*, 1987). Moreover, following inoculation of transgenic plants with the challenge virus, fewer lesions appeared and virus accumulation was substantially lower than that seen with control plants.

Similar results have been obtained with another RNA virus, alfalfa mosaic virus (AlMV) (see Fig. 7.4) (Tumer *et al.*, 1987). AlMV is unrelated to TMV having a different morphology, genome structure and gene expression strategy. Hence, the observation that symptoms of AlMV infection were reduced following inoculation of plants expressing the AlMV coat protein suggests that this method of suppressing symptoms of infection might be applicable to a variety of RNA viruses. Once again symptom suppression was correlated with a reduction in the levels of virus accumulating in the inoculated tissue, although suppression of symptoms could not be overcome by inoculating with increasing amounts of RNA.

Fig. 7.4 Suppression of AlMV symptoms in tobacco plants engineered to over-express the AlMV coat protein. The plant on the left is normal whereas the plant on the right has been engineered to contain the AlMV coat protein. The plants were photographed 4 weeks after inoculation with AlMV. (Reproduced with permission from Tumer *et al.*, 1987.)

The mechanism by which cross protection acts is not known although a clue might be provided by the observation that, with TMV, protection is largely overcome by inoculation with viral RNA instead of the virus particle and it appears that it is the presence specifically of the coat protein gene in the transgenic plants which is required for cross protection to take place. This suggests that protection may result from the coat protein interfering with the entry of the virus into the cell or the uncoating of the viral RNA.

The second approach to engineering the suppression of symptoms of viral infection has been introduced in Chapter 6 where the insertion of a cDNA representing a satellite RNA of CMV into the plant genome was described. This work has been extended to show that when tomato aspermy virus (TAV), which is closely related to CMV, is inoculated onto plants which have been engineered to express CMV satellite RNA, there is an induction of satellite RNA synthesis which is accompanied by the suppression of symptoms of viral infection (Harrison *et al.*, 1987). Following inoculation of the transgenic plants with TAV, in contrast to CMV, there appears to be no inhibition of replication of genomic RNA. Hence, although the reduction of symptoms following inoculation with CMV might be related to reduction in the synthesis of viral particles, this is not the case with TAV. The reduction in symptoms appears to be relatively specific as there was no effect seen following inoculation of transgenic plants with other tripartite RNA viruses, with viruses that contain a satellite RNA or with other cucumoviruses.

Similar results have been obtained with inoculating plants engineered to contain the satellite of tobacco ringspot virus (ToRV) with ToRV (Gerlach *et al.*, 1987). Once again suppression of symptoms was accompanied by a decrease in the levels of genomic RNA replicating in the inoculated tissue.

This strategy for producing a reduction in viral symptoms has the advantage over the use of the viral coat protein in that the transgenic tissue does not synthesize constitutively high levels of viral coat protein, rather low levels of RNA that become amplified upon infection. Moreover, protection is not overcome by increasing the amount of RNA in the inoculum. On the other hand, while the satellite RNA might serve to alleviate symptoms on one plant, it might prove more virulent if transferred to a new host or might mutate to become so. This can be avoided if the satellite RNA can be modified so that it is no longer transmitted from the transgenic plant.

Another potential method of engineering resistance may be to engineer the plant to produce an antisense RNA which will interfere with viral nucleic acid replication during the initial stages of infection. In several experimental systems it has been found that antisense RNA can serve to modulate gene expression by hybridizing to sense RNA, disrupting its function. This approach may be applicable to viruses with either an RNA or a DNA genome. Hence, the idea would be to use a sequence of DNA (or cDNA) to direct the expression of an antisense RNA which would then hybridize to a region of viral RNA or DNA in such a manner as to interrupt its replication. A possible drawback of this approach is that there needs to be an excess of antisense RNA in order for it to work efficiently, and it needs to hybridize to a region of viral sequence which is important to replication so as to disrupt the viral replication cycle.

Engineering Plants to Overproduce Foreign Proteins

In addition to the examples cited there is of course the possibility of using engineered plants to synthesize large amounts of a particular polypeptide. For example, engineering potato tubers to synthesize insulin or human growth hormone is feasible. However, at this point the idea is likely to leave the realms of science and enter the realms of economics with the question of whether it is a financially worthwhile proposition. It is all very well to arrange for a field of oilseed rape to synthesize blood clotting factor but what will the costs of extraction be, compared with, say, mammalian cell, yeast or indeed plant cell suspension culture? Will it be free of contaminants and of high enough purity for use as a pharmaceutical? With the idea of using plants to overproduce a foreign polypeptide, possibly one of the best approaches would be to engineer a portion of the plant which is usually discarded, say, the foliage of sugar beet, to accumulate a protein that enhances its nutritional content as an animal food, or improves it as a substrate in compost making. This approach circumvents all the difficulties of isolation and purification of a foreign protein yet produces a value added quality to a by-product that might not be used normally.

Conclusions

From this discussion we can see that we are still a long way from being able to engineer some of the more agronomically useful traits into crop plants. Nevertheless, advances in the past few years have been great and based squarely on our growing knowledge of how to transfer genes into plants, arrange for their correct expression and targeting of their protein products to the correct cellular compartment. This provides a clear example of how advances in fundamental research have gone hand in hand with the development of practical applications. Already many of the engineered plant lines that have been described in this chapter are undergoing field trials. One of the novelties of genetic engineering is that it can be carried out on plant lines that are already commercially successful. This could potentially decrease the time it takes for the engineered plant to become a commercial product; nevertheless, extensive trials still need to be carried out to test the stability and performance of the engineered trait as well as what effect it may have on crop yields. Alternatively, where engineered traits have not been introduced into commercial varieties the transformed plants may provide useful material for use in conventional breeding programmes. As shall be seen in Chapter 8, these advances are likely to continue and gene transfer technology may yet allow us to dissect the plant genome further so that we can begin to address the problem of how to isolate the complex families of genes that are involved in determining the traits that are important to the plant breeder.

References

Beversdorf, W.D., Weiss-Lerman, J., Erickson, L.R. and Souza-Plachada, Z. (1980). Transfer of cytoplasmic inherited triazine resistance from birds rape to cultivated oilseed rape (*Brassica campestris* and *Brassica napus*). *Can. J. Genet. Cytol.* **22**, pp. 167–172.

Chaleff, R.D. and Mauvais, M. (1984). 'Acetolactate synthase: site of action of 2-S-sulfonylurea herbicides', *Science* **224**, pp. 1443–1445.

Comai, L., Facciotti, D., Hiatt, W.R., Thompson, G., Rose, R.E. and Stalker, D.M. (1985). 'Expression in plants of a mutant *aroA* gene from *Salmonella typhimurium* confers tolerance to glyphosate', *Nature* **317**, pp. 741–744.

Dean, D.H. (1984). 'Biochemical genetics of the bacterial insect-control agent *Bacillus thuringiensis*: Basic principles and prospects for genetic engineering', *Biotechnology and Genetic Engineering Reviews* **2**, pp. 341–363.

DeBlock, M., Botterman, J., Vandewiele, M., Dockx, J., Thoen, C., Gossele, V., Rao Movva, N., Thompson, C., Van Montagu, M. and Leemans, J. (1987). 'Engineering herbicide resistance into plants by expression of a detoxifying enzyme', *EMBO J.* **6**, pp. 2513–2518.

della-Chioppa, G., Bauer, S.C., Klein, B.K., Shah, D.M., Fraley, R.T. and Kishore, G.M. (1986). 'Translocation of the precursor of 5-enolpyruvylshikimate-3-phosphate synthase into chloroplasts of higher plants *in vitro*', *Proc. Nat. Acad. Sci. USA* **83**, pp. 6873–6877.

della-Chioppa, G., Bauer, S.C., Taylor, M.L., Rochester, D.E., Klein, B.K., Shah, D.M., Fraley, R.T. and Kishore, G.M. (1987). 'Targeting a herbicide-resistant enzyme from *Escherichia coli* to chloroplasts of higher plants', *Bio/Technology* **5**, pp. 579–588.

Fillatti, J., Kiser, J., Rose, R. and Comai, L. (1987a). 'Efficient transfer of a glyphosate tolerance gene into tomato using a binary *Agrobacterium tumefaciens* vectors', *Bio/Technology* **5**, pp. 726–730.

Fillatti, J., Sellmer, J., McCown, B., Haissig, B. and Comai, L. (1987b). '*Agrobacterium* mediated transformation and regeneration of *Populus*', *Mol. Gen. Genet.* **206**, pp. 192–199.

Fischhoff, D.A., Bowdish, K.S., Perlak, F.J., Marrone, P.G., McCormick, S.M., Niedermeyer, J.G., Dean, D.A., Kusano-Kretzmer, K., Mayer, E.J., Rochester, D.E., Rogers, S.G. and Fraley, R.T. (1987). 'Insect tolerant tomato plants', *Bio/Technology* **5**, pp. 807–813.

Fulton, R.W. (1986). 'Practises and precautions in the use of cross protection for plant virus disease control', *Ann. Rev. Phytopath.* **24**, pp. 67–81.

Gerlach, W.L., Llewellyn, D. and Haseloff, J. (1987). 'Construction of a plant disease resistance gene from the satellite RNA of tobacco ringspot virus', *Nature* **328**, pp. 802-805.

Hamill, J.D., Parr, A.J., Rhodes, M.J.C., Robins, R.J. and Walton, N.J. (1987). 'New routes to plant secondary products', *Bio/Technology* **5**, pp. 800–806.

Harrison, B.D., Mayo, M.A. and Baulcombe, D.C. (1987). 'Virus resistance in transgenic plants that express cucumber mosaic virus satellite RNA', *Nature* **328**, pp. 799–805.

Hilder, V.A., Gatehouse, A.M.R., Sheerman, S.E., Barker, R.F. and Boulter, D. (1987). 'A novel mechanism of insect resistance engineered into tobacco', *Nature* **330**, pp. 160–163.

Mazur, B.J., Falco, S.C., Knowlton, S. and Smith, J.K. (1987). Acetolactate synthase, the target enzyme of the sulfonylurea herbicides in *Plant Molecular Biology*, Eds. von Wettstein, D., and Chua, N-H. pp. 339-350. New York, Plenum.

Moreland, D.E. (1980). 'Mechanism of action of herbicides', *Ann. Rev. Plant Physiol.* **31**, pp. 597–639.

Nelson, R.S., Powell Abel, P. and Beachy, R.N. (1987). 'Lesions, and virus accumulation in inoculated transgenic tobacco plants expressing the coat protein gene of tobacco mosaic virus', *Virology* **158**, pp. 126–132.

Powell Abel, P., Nelson, R.S., De, B., Hoffman, N., Fraley, R.T. and Beachy, R.N. (1986). 'Delay of disease development in transgenic plants that express tobacco mosaic virus coat protein gene', *Science* **232**, pp. 738–743.

Shah, D., Hironaka, C.M., Wiegand, R.C., Harding, E.I., Krivi, G.G. and Tiemeier, D.C. (1986a). 'Structural analysis of a maize gene coding for glutathione-S-transferase involved in herbicide detoxification', *Plant Mol. Biol.* **6**, pp. 203–211.

Shah, D., Horsch, R.B., Klee, H.J., Kishore, G.M., Winter, J.A., Tumer, N., Hironaka, C.M., Sanders, P.R., Gasser, C.S., Aykent, S., Siegel, N.R., Rogers, S.G. and Fraley, R.T. (1986b). 'Engineering herbicide tolerance in transgenic plants', *Science* **233**, pp. 478–481.

Thompson, C.J., Rao Movva, N., Tizard, R., Crameri, R., Davies, J.E., Lauwereys, M. and Botterman, J. (1987). 'Characterisation of the herbicide-resistance gene *bar* from *Streptomyces hygroscopicus*', *EMBO J.* **6**, pp. 2519–2523.

Tumer, N., O'Connell, K.M., Nelson, R.S., Sanders, P.R., Beachy, R.N., Fraley, R.T. and Shah, D.M. (1987). 'Expression of alfalfa mosaic virus coat protein confers cross-protection in transgenic tobacco and tomato plants', *EMBO J.* **6**, pp. 1181–1188.

Vaeck, M., Reynaets, A., Hofte, H., Jansens, S., De Beuckeleer, M., Dean, C., Zabeau, M., Van Montagu, M. and Leemans, J. (1987). 'Transgenic plants protected from insect attack', *Nature* **328**, pp. 33–37.

Chapter 8

Future Directions in Plant Transformation

Introduction

The transfer of foreign genes into plants followed by their correct expression in a tissue-specific, developmentally regulated manner is now a matter of routine practice for a small, but growing number of species. This has enabled us to begin to understand the molecular basis of gene expression in general and in particular the functional significance of the sequences flanking genes. Furthermore, it allows the investigation of the function of individual gene products during the life cycle of the plant. At the practical level, following recent governmental approval, genetically engineered plants are now undergoing field trials in several countries so that their engineered traits, for example herbicide resistance, can be tested under field conditions. Indeed, the protagonists of engineering herbicide tolerance into plants confidently predict that this research is rapidly approaching fruition and that transgenic plants will become commercially available in a relatively short time. Hence, we may be approaching the stage where the initial phase in establishing gene transfer as a practical tool in plant breeding has come to an end. If this is the case, the next stage will be one of consolidation where the initial findings are refined and improved. Doubtless more genes will be transferred to the plant genome to modify the biochemistry of the cell, but if gene transfer is to be used widely as a tool in crop improvement the molecular biologists will need to begin to tackle the question of engineering more subtle traits rather than simple resistance to herbicides and plant pathogens. Following the discussion of how all this has come about it might be worthwhile to pause and consider what has been achieved, where limitations may lie with current technology, and what future directions might be open for exploration using gene transfer in plants.

Further Development of Transformation Strategies and *Agrobacterium*-based Vectors

Techniques for plant transformation have been developed for use on a variety of plant species. With naked DNA uptake, high frequencies of transformation can be obtained by electroporation and/or incubation in the presence of polyethylene glycol. Using DNA contained on Ti plasmid-based vectors the technique of protoplast co-cultivation has been largely superseded by explant inoculation, e.g. with leaf discs. Nevertheless, whichever method is used to insert DNA into plant cells one important factor remains the same: the cells into which the foreign DNA is introduced must be able to divide and regenerate into a whole plant so that the DNA is inserted into the germline. It is probable that the list of plants that are transformed by these methods will continue to grow and begin to include some of the world's more important crop plants. Recent reports from a number of laboratories of the regeneration of both rice and maize plants from isolated protoplasts signal that it is only a matter of time until regenerated transgenic plants from these species expressing NPTII or GUS are reported in the scientific press.

At the same time, it is likely that vectors based on the Ti plasmid will continue to improve, becoming increasingly versatile as genetic markers continue to be developed. This will go hand in hand with the further clarification of how *Agrobacterium* transfers the T-DNA to the plant cell. We have seen that there is now compulsive evidence that *Agrobacterium* can transfer its DNA to monocotyledonous cells and that the methods used previously, such as monitoring tumour formation or the synthesis of opines, which suggested that this was not the case, were generally not sensitive enough to detect conclusively the transfer of the T-DNA. Hence, although efficiencies in transfer may differ from that seen with dicotyledonous plants, there is no potential barrier to *Agrobacterium* transferring DNA into monocotyledonous plant cells provided that the experimental system is carefully controlled. In addition, isolation of *Agrobacterium* strains that have wide host ranges and with different patterns of virulence will also be of benefit in establishing transformation protocols for other species of plants. Thus, it is the limitations of plant regeneration rather than the inability of currently available vectors to be transferred or expressed in these cells that is likely to remain the major barrier in producing transgenic monocotyledonous plants.

Tissue Culture-free Approach to Transformation

Where the barrier to transformation is our inability to regenerate plants from single cells this difficulty might be circumvented by other methods currently considered as somewhat unorthodox. An example of this would be the bombardment of tissue explants or partially organized tissue which can be regenerated, such as embryoids, by microprojectiles coated with DNA (see Chapter 5). Indeed the need to use tissue culture at all has been called into question with the report of the transformation of rye by the injection of DNA into the developing flower (De la

Pena *et al.*, 1987). In order to do this, developing tillers of rye were injected with plasmid DNA containing an NPTII gene linked to the *nos* promoter. Located within the tillers are the archesporial cells which give rise to pollen as a result of meiosis in the developing pollen sac. Previous work had shown that two weeks before the first meiotic metaphase the archesporial cells are permeable to caffeine and colchicine and this led to the question of whether the cells could also take up larger molecules such as DNA. Plants which had been injected with the plasmid DNA were crossed and the resultant seeds were tested for their ability to germinate in the presence of kanamycin. Of 3023 seedlings tested, 7 remained unbleached in the presence of the antibiotic and of these, 2 contained NPTII activity. These seedlings contained NPTII specific DNA within their nuclear DNA and one plant which was kanamycin resistant, yet contained no detectable NPTII activity, contained plasmid specific DNA in the nuclear genome that had been rearranged, suggesting that modifications of the DNA took place prior to the integration, reminiscent to that seen with naked DNA uptake into the plant cell. However, the exact mechanism of how the DNA enters the plant cell remains unknown. This work was the first report of a transgenic gramineous monocotyle-donous plant and the technique, although dependent on the precise timing and the location of the injection, might be applicable to other major cereal plants. Moreover, this result demonstrates that strategies of transformation in the absence of tissue culture are indeed feasible and it would seem that efforts to introduce DNA into plants have come a complete circle from the initial attempts where DNA was applied to plant tissue in the 1970s!

Marker Genes Used to Isolate Plant DNA Sequences

Of course the key to the success of all this work has been the development of genetic markers that can be used to show that the foreign DNA is integrated into the plant genome and is being expressed. The number of markers available is likely to increase and the use of markers expand from simply being used to demonstrate that a plant contains a piece of foreign DNA. As discussed in Chapter 3, markers allowing negative selection are likely to be developed in order to study gene inactivation and refinement of the *lux* marker system may allow us to study gene expression in plant cells in a non-destructive manner which would be of great benefit to the study of developmental gene expression. We have already seen that different markers have been used in promoter analysis and they can also be used to select for sequences of genomic DNA that can act as promoter sequences. There are two ways in which this can be carried out. One approach is to clone random fragments of the plant genome upstream from a promoterless NPTII gene and select for their ability to confer kanamycin resistance following transfer to the plant genome (Herman *et al.*, 1986). In this case a binary transformation vector was constructed which contained a NPTII gene with a unique BglII restriction site 4 bp upstream from the initiation codon of the coding sequence. Random fragments of the plant genome were inserted into this site and pools of recombinant plasmids transferred into *Agrobacterium* which was then used to

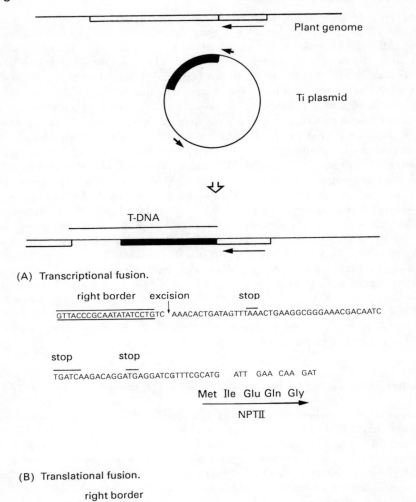

Fig. 8.1 The principle of using the T-DNA to isolate plant promoters. The transformation vector used is engineered so that the right T-DNA border is located next to an NPTII gene. Following transfer to the plant cell any transformants which contain T-DNA located next to a plant promoter will be selected for by conferring resistance to kanamycin. □ = plant gene; long arrows indicate plant promoter sequence; short arrows indicate T-DNA border repeats, ■ = NPTII gene. (A) A construct for transcriptional fusions; (B) a construct for translational fusions.

transform isolated protoplasts. Selection for kanamycin resistance was applied at the time of callus formation and a number of kanamycin-resistant calli and plants were obtained. The other approach is to use the T-DNA as a molecular tag to isolate promoter sequences directly from plants. The idea here is to construct a promoterless marker gene and place it downstream from the right border sequence of the T-DNA and transfer it to the plant genome. If the sequence has been inserted downstream from a plant promoter, the marker will be expressed by read-through from the plant promoter into the T-DNA thus forming a transcriptional fusion. This has been accomplished using the NPTII gene (Teeri *et al.*, 1986). In order to do this the NPTII coding sequence was placed 71 bp downstream from the normal junction site of the plant DNA and the T-DNA just within the right border sequence (see Fig. 8.1a). The sequence between the border repeat and the NPTII gene contained two translational stop codons so that read-through from the plant genomic sequences would not form an active translational fusion with NPTII. Hence, when this construct is inserted into the plant genome, any differences in the activity of the NPTII would not result from differences in the stability or the activity of the chimeric protein. The construct was inserted into plant cells by co-cultivation of the *Agrobacterium* with tobacco protoplasts and selection for transformants was applied at the callus level. Using both techniques, a variety of transgenic kanamycin-resistant plants were obtained and it is significant that the majority of the promoters that were identified were stem-specific in their pattern of expression. This may be a reflection of the selection being applied at the level of callus formation with only promoters being expressed at this stage being selected for. This would suggest that those promoters active normally in callus also function in the stem. The frequency of isolation of a particular type of promoter could be changed either by applying the selection at a different stage of development or by using a different selectable marker to select for transformants prior to screening for different promoter activities at different stages of development.

Once the promoter has been tagged with the T-DNA it can be isolated. Normally this would involve preparing a genomic library from the nuclear DNA of the transgenic tissue and then screening for recombinants that contain the T-DNA and flanking plant sequences. An alternative approach is to make a T-DNA vector which contains an *E. coli* plasmid origin of replication and a selectable marker functional in bacteria. The DNA can then be isolated from the transgenic tissue, cut with a restriction enzyme that cleaves only in the plant DNA flanking the inserted T-DNA, re-ligated and transformed directly into *E. coli* and rescued by selection for the bacterial marker present in the T-DNA (Koncz *et al.*, 1984).

Organelle Transformation

As described in Chapter 2 it has been known for some time that the T-DNA is transferred to the plant nuclear genome. However, several important agronomic traits, such as herbicide resistance or cytoplasmic male sterility, are encoded by the chloroplast or mitochondria so there is a great amount of interest in being able to engineer the DNA of these organelles. Despite this, as yet there is no reliable method for inserting DNA into the chloroplast or mitochondrial genomes although there has been one report of T-DNA becoming integrated into the plastid genome. This was a very rare event which was detected by the NPTII activity encoded by the transferred DNA being inherited in a non-Mendelian manner (DeBlock *et al.*, 1985). Organelle transformation is complicated because not only does the foreign DNA have to enter the organelle but also the plant cell may contain many individual organelles with each containing multiple genomes. This difficulty is compounded by the finding that plant mitochondrial genes appear to obey a slightly different genetic code to that of other genetic systems (see for review Leaver *et al.*, 1983) which means that any foreign gene to be expressed in the mitochondrion would need to be modified accordingly. Hence, at the moment the only feasible way of engineering a protein to be enzymatically active in the chloroplast or the mitochrondria is to transfer the gene encoding it to the plant nuclear genome and arrange for it to be fused to a transit peptide so that it can be post-translationally transferred to the organelle in question.

Transfer of More Than One Gene to Plant Cells

To date, it has been the norm for only single plant genes to be transferred to the plant genome, although in the future it may become increasingly important to transfer more than one gene and ensure that they are expressed in a coordinated manner, particularly if novel biochemical pathways are to be engineered into plants. So far, there have only been a few examples of sequences of DNA containing more than one gene being transferred to the plant genome where they are expressed in a coordinated manner, the T-DNA, viral genomes, the two genes of the *Vibrio harveyii lux* operon and the four soybean genes which were transferred on a single fragment of DNA to the tobacco genome (see Chapter 6). In the case of the T-DNA and the *lux* operon two or three genes are naturally, or are engineered to be, coordinately expressed so that the plant cell contains a new biochemical pathway. Engineering novel biochemical pathways into plants will involve the transfer of not only the individual genes but also their upstream controlling elements. Should the pathway be the same in the new host plant as in the plant that serves as the source of the genes the upstream regions of the genes are likely to be capable of directing the correct coordinated expression in their new environment. If, however, the genes are from a non-plant donor, or they are to be expressed in a novel developmental manner, then the upstream controlling regions which will direct this expression need to be isolated and fused to the genes in question.

Plant Promoters and the Role of Individual Gene Products in Development

Until now, the promoter sequences that have been isolated have generally been those that direct the expression of abundant or developmentally regulated polypeptides. Increasingly there will be a requirement for promoter sequences with a more subtle control on gene expression. Bearing this in mind, it is nevertheless clear that gene transfer will allow us to study the functions of gene products in novel environments in order, not only to assess their role but also to investigate, via *in vitro* mutagenesis, which portions of the protein are important to their function. An example of this has already been provided by the transfer of the T-DNA genes encoding the enzymes that synthesize plant hormones to potato and assessing the role of endogenously synthesized cytokinins on potato tuber and plant development (Ooms and Lenton 1985). This type of work has great potential to the plant biologist interested in the role of plant growth substances in development because it offers the opportunity of being able to induce the synthesis of predictable amounts of auxin or cytokinin and study the effects on the plant rather than having to rely on observing the effects of these compounds following exogenous application. At the moment we have the opportunity to induce the over-production of plant growth substances in a tissue-specific manner; however, the promoters that are currently available have two drawbacks: first, they can be 'leaky', directing very low levels of transcription in the absence of induction rather than being totally inactive; second, the factor that induces their expression is also likely to induce the transcription of a variety of other genes in the plant cell so that it is difficult to be able to assess the effect of the expression of a single induced gene. Hence, the isolation, or the construction, of a promoter which can be induced uniquely to carry out transcription would be of great benefit.

Gene Inactivation

One of the difficulties in attempting to study the effect of a particular gene in a plant by transferring it into a new host is the presence of the endogenous homologous protein encoded by the host plant which may interfere with the results. This difficulty is compounded further by the observation that many plant genes are members of multigene families. At present there is no method available to allow gene inactivation in a manner similar to that enjoyed by molecular biologists working with yeast or some mammalian cell lines where gene inactivation can result from either homologous recombination or a mutation arising from the incorrect repair of mismatches formed by heteroduplexes between the host sequence and the incoming DNA. This is because of two reasons: first, there is an absence of useful genetic markers based on genes present in the nuclear genome; second, there is a relative lack of sensitivity of assays for gene inactivation. Nevertheless, we have already seen that recombination can take place between incoming DNA following naked DNA uptake in an analogous manner to that seen in mammalian cell systems and with appropriate markers and selection procedures gene targeting might be achieved by microinjection, electroporation or PEG-mediated uptake.

Another method of inactivating a specific gene sequence is to engineer the plant to over-express an antisense RNA corresponding to the sequence in question. As was seen in Chapter 5, this was achieved in protoplasts that have been electroporated with DNA constructed so as to synthesize sense and antisense RNA in a transient assay system. Recently it has also been achieved in transgenic tobacco plants (Rothstein *et al.*, 1987). In this case a construct consisting of an antisense *nos* gene fused to the 35S RNA promoter and poly(A) addition site was found to inhibit expression of a normal *nos* gene that had been previously introduced into the genome of the host plant. The inhibition in expression is a result of a decrease in the amounts of sense *nos* mRNA and appears to be stably inherited indicating that this indeed might be a method of modifying the phenotype of the plant.

The difficulty of the expression of endogenous plant genes can be partially overcome by using mutant plants or cell lines derived from tissue culture. Variants can be obtained experimentally by treatment of the cells with mutagens such as N-ethyl-N-nitrosourea or may arise spontaneously by somaclonal variation (for review see Maliga 1984). However, care must be taken, by the investigation of sexual transmission of the altered phenotype, that the change is due to genetic mutation and not the result of epigenetic factors. Several mutants have been well characterized and these include mutants in amino acid, nitrate and carbon metabolism. Another method of inactivating specific gene sequences is insertional mutagenesis and with this in mind attention has focused increasingly on plant transposable elements as agents to carry this out.

Gene Tagging by Transposons

The pioneering work of McClintock in the 1930s and 1940s led to the acceptance of the idea that plant transposable elements are discrete pieces of DNA that can move around the plant genome and in so doing disrupt gene function by insertion into the gene itself (either in an exon or intron sequence) or the regions surrounding the gene that are important in controlling its expression. This has allowed not only the generation of novel phenotypes but also the isolation of the genes which have been mutated by using transposon-specific hybridization probes to identify the mutant genes in genomic libraries. By analogy with work with *Drosophila*-transposable elements this has come to be known as 'gene tagging'. Gene tagging has played an important role in plant molecular biology, being used to isolate a variety of genes, to study the manner in which they are development-ally expressed and to find what role they play. However, this system is currently limited to the relatively few species in which transposons have been characterized and the mutant phenotypes which have been easily identified. These limitations can, however, be partially overcome by the finding that transposable elements can work in a novel plant host following transfer to the genome mediated by the Ti plasmid. An example of this involves the *Ac/Ds* transposable element system from maize (Baker *et al.*, 1986). Briefly, this system comprises of two elements, *Ac* (activator) which is an autonomous element, able to transpose by itself, and *Ds* (dissociator) which is non-autonomous and can only transpose when *Ac* is present in the genome. *Ac* is a 4.5 kb sequence of DNA that contains three open reading frames, has 11 bp terminal repeat sequences and upon insertion into the genome generates an 8 bp duplication of plant DNA at the site of insertion. *Ds* elements are a heterogeneous group of elements but all have similar or identical 11 bp terminal

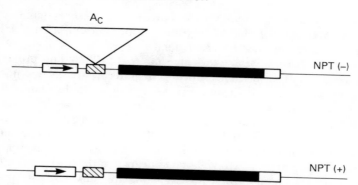

Fig. 8.2 Phenotypic assay of transposition. The *Ac* element inserted in the untranslated leader sequence between the promoter and the NPTII coding region results in sensitivity to kanamycin. Transposition of the *Ac* element results in kanamycin resistance. Block containing arrow represents the promoter of gene 1 from the octopine T_R DNA; ▨ = the *wx* sequence in which the Ac element is contained; ■ = the NPTII coding region; □ = the *ocs* poly (A) addition site.

repeats to those of *Ac*. Some *Ds* elements have arisen directly from *Ac* by mutation. In order to test transposition of *Ac* in tobacco, three different constructs were transferred to the genome using a Ti plasmid-based vector: an *Ac* element inserted into the 'waxy' (*wx*) locus, a *Ds* element inserted within a comparable *wx* fragment and the *wx* locus alone. Southern blot analysis of DNA from the transformed cells showed that in transgenic tissue, containing the *wx* locus alone or the *Ds* element in *wx*, the insertion was stable, whereas in the transgenic tissue containing the *Ac* element in *wx*, the *Ac* element had transposed from the *wx* locus and inserted into other sites in the genome in 50% of the cases studied. The *wx* locus from which the *Ac* element had transposed was cloned from tobacco and sequence analysis showed that all that remained to suggest that the *Ac* element had been present was an 8 bp duplication of DNA which was identical to the transposon 'foot print' that remains in the maize genome after *Ac* has transposed from its site of insertion. Thus it would seem that the mechanism of transposition in maize and tobacco is the same and that it was the *Ac* element alone which was able to catalyse transposition. In the initial work, transposition could only be assessed by Southern blot analysis of the genomic DNA from the transgenic tissue; however, subsequently a phenotypic marker of transposition has been developed (Baker *et al.*, 1987). In this case the *Ac* element, flanked by 60 bp of the *wx* locus, was inserted into the untranslated leader sequence between the gene 1 promoter of octopine T_R DNA and the NPTII gene and the construct was transferred to the tobacco genome. The presence of the *Ac* element in the untranslated leader sequence of the NPTII gene disrupts its expression but this can be restored by the transposition of the *Ac* element as the remaining 60 bp of *wx* DNA does not interfere with expression of the NPTII gene (see Fig. 8.2). This then allows a phenotypic assay for transposition of the *Ac* element following transfer to a new plant host simply by screening for resistance to kanamycin and provides a marker system to assess the effect of mutations of the *Ac* element on transposition. Transposition of *Ac* has been observed also in carrot and *Arabidopsis* following transfer on a binary plasmid vector from *Agrobacterium rhizogenes* (Van Sluys *et al.*, 1987). This provides us an opportunity to use a transposable element system to carry out *in vivo* mutagenesis of the plant genome following transfer of the *Ac* element to the plant cell on a Ti plasmid vector. However, this approach will not be without its complications because the transposable elements are likely to insert into a great many sites in the genome. On one hand, this may be useful allowing an increased probability of mutating a particular target gene, while on the other it will complicate the isolation of the mutant gene. Moreover, it appears that the transposable elements continue to transpose in transgenic tissue making any mutants potentially unstable.

The Use of T-DNA as a Mutagen

The difficulties in using transposable elements in gene tagging can be partially overcome by using T-DNA as a mutagen because it is likely to insert into one site in the genome and is stable and unlikely to revert. This approach has been demonstrated by the transformation of haploid and diploid *N. plumbaginifolia* with Ti plasmid-based vectors. Haploid cells were used because recessive mutations will not be concealed by the presence of a wild-type allele and diploid plants can be used to recover recessive lethal mutations which will be detectable after selfing. In this case a vector with a promoterless NPTII gene placed 30 bp away from the right border sequence and containing no intervening translational stop signals was used so that read-through from the plant genome results in a translational fusion (Fig. 8.1b) (Andre *et al.*, 1986). Using the T-DNA as an insertional mutagen in this manner allows not only the isolation of mutant tobacco plants but also yields information regarding the expression of the mutated gene. Obviously this type of experiment can also be carried out with a marker, such as GUS or *lux*, allowing histochemical or optical detection of promoter activity. The beauty of using this type of marker in a study of this sort is the potential for the enzyme activity to be detected very precisely, at the cellular level if necessary, and also possibly in a non-destructive manner. Nevertheless, whether the T-DNA or a transposible element is used as a mutagen one key factor remains: the possible inability to select for, or observe, a mutant phenotype worth further investigation.

Gene Rescue from the Plant Genome

Where mutant plant cell lines are available, the technique of complementation of the phenotype as a method of gene isolation or the functional analysis of the gene itself is possible. We have already seen that complementation of a mutant cell line and a whole plant has been achieved following Ti plasmid-mediated gene transfer (Chapters 3 and 6). However, in these examples, complementation was achieved using previously characterized genes. One of the goals of experiments involving complementation is the rescue of specific genes from the plant genome by shotgun cloning followed by transfer to a mutant cell line. Gene rescue by complementation has been applied routinely to both procaryotic and eucaryotic systems to isolate genes; however, when applied to plants the technique is fraught with practical difficulties including the following.

(*a*) *Availability of mutant cell lines*. Although these are available from a variety of plant species the numbers are still relatively small and the mutations themselves might not be of practical interest.

(*b*) *Source of plant genes*. The initial gene library representing the source of DNA needs to be constructed so that it represents a large part of the plant genome ($>95\%$). This demands that the genome is relatively small so that the number of clones necessary to ensure this is not too large. Using current technology this essentially reduces the source of the DNA to being a plant with a very small genome such as *Arabidopsis* (7×10^4 kb).

(*c*) *Transformation vector*. The library needs to be constructed in, or transferred to, a plant transformation vector for passage to the plant cell. The cloned sequences must be stable in both *E. coli*, where all the manipulations are carried out, and *Agrobacterium* into which the library must be efficiently transferred prior to passage into the plant cell. All members of the library should be transferred to the plant genome with equal efficiency regardless of size.

(*d*) *The method of selection*. In order to be certain that a particular genomic sequence has been transferred to the plant genome several hundred thousand cells need to be screened. For practical purposes screening can most conveniently be carried out at the callus level but this requires that the gene to be isolated is expressed at this developmental stage.

As yet the rescue of a plant gene from a shotgun library of a plant genome by complementation has not been reported. However, there have been several studies of the feasibility of this type of approach to gene isolation. The stability of a shotgun library of a plant genome in both *E. coli* and *Agrobacterium* has been studied in some detail (Simoens *et al.*, 1987). In these experiments a genomic library from *Arabidopsis thaliana* was constructed in a binary plant transformation vector, pC22. This vector contains several components, an origin of replication from the Ri plasmid and a streptomycin/spectinomycin marker to allow maintenance in *Agrobacterium*, the origin of replication from pBR322 and the ampicillin resistance gene allowing manipulation in *E. coli*, the right and left T-DNA border sequences flanking the cloning site, a *cos* site and a marker for selection of transgenic plants. It was found that approximately 60% of the clones containing plant DNA were unstable in the bacteria and that the instability was independent of the size of the insert DNA suggesting the instability might be a feature of the plant DNA.

The efficiency of transfer of DNA from *Agrobacterium* to the plant cell has been studied in detail by investigating the frequency of transfer of a small genomic library to the plant cell (Prosen and Simpson 1987). In order to do this the genome of bacteriophage T7 was digested to yield 10 fragments and these were cloned into a binary plant transformation vector. This library was then transferred to the tobacco genome and transgenic plants analysed for the presence of T7 DNA integrated into the genome. Southern blot analysis of the DNA from 66 transgenic plants demonstrated that each of the 10 library members were present in at least 2 plants. The majority of plants contained a single library member although some contained 2 or more with one containing 4, suggesting that transformation by more than one bacterial cell had taken place. However, 20% of all the T-DNA copies were rearranged or incomplete. This work indicates that gene rescue from the plant genome by the complementation of shotgun clones will be by no means easy. The possible instability of plant DNA in bacteria will, at best, mean that the number of clones to be analysed will need to be increased to be certain that all sequences are represented and, at worst, that some sequences will not be represented at all because of their inherent instability. This may be partially overcome if multiple inserts of plant DNA can be made into the cloning vector or if all the manipulations involved in the construction of the shotgun library can be carried out in *Agrobacterium*. In addition, the number of transformation events required to transfer a complete gene library into plants might be reduced if the conditions can be manipulated so as to allow the transfer of multiple sequences of DNA into the plant genome.

An example of a gene rescued from a plant genome has been provided by the rescue of an NPTII gene from the genome of *Arabidopsis* (Klee *et al.*, 1987). In this case the NPTII gene flanked by the promoter and termination signals of *nos*, the *nos* gene and a gene which encodes resistance in bacteria to spectino-mycin/streptomycin was introduced into *Arabidopsis*. A genomic library was constructed from the transgenic tissue in a vector that contained a broad host range origin of replication (*oriV*), a sequence directing the transfer from *E. coli* to *Agrobacterium* (*oriT*), a bacterial selectable marker sequence and the T-DNA border sequences flanking not only the site into which the plant DNA is inserted but also a bacteriophage lambda *cos* sequence (see Fig. 8.3). The last is important for very large segments of plant DNA to be recovered via *in vitro* packaging. Recombinants containing the T-DNA insert flanked by *Arabidopsis* DNA were recovered in the bacteria by selecting for spectinomycin resistance and were present at a frequency of 1 in 3000. The complete library was transferred to *Agrobacterium* and used to transform *Petunia* leaf discs with transformants being selected for by their resistance to kanamycin. A number of calli containing the original DNA were obtained. This work demonstrates that the overall strategy of gene rescue from one plant genome into another may be feasible; however, it needs to be borne in mind that a powerful dominant selectable marker was used and many useful genes might not be so easily selected for.

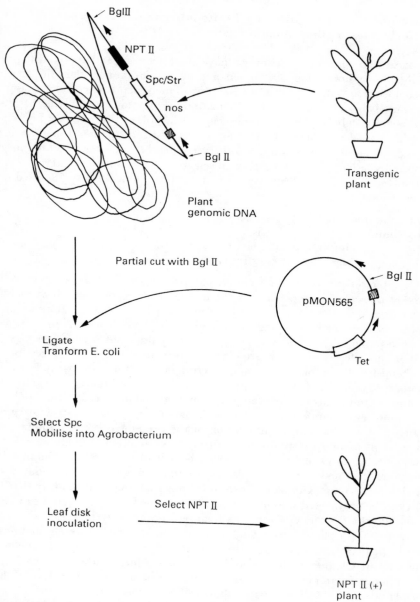

Fig. 8.3 Schematic representation of gene rescue. Total plant DNA containing the NPTII gene is cloned into a binary vector, mobilized into *Agrobacterium* and transferred to the plant genome conferring kanamycin resistance. BglII restriction sites are shown; Spc refers to the bacterial gene encoding spectinomycin resistance; ▨ = a *cos* site; the bold arrows represent the T-DNA border sequences.

The Potential for Engineering Traits other than Resistance to Herbicides, Pathogens and Pests

It is clear that while these techniques have great potential and are likely to prove invaluable in improving our understanding of plant molecular biology and biochemistry as well as assist in the isolation of more plant genes, these examples have largely been based on model systems and the question remains whether these approaches will be applicable to engineering some of the traits of most interest to the plant breeder. Whereas some traits, as discussed in Chapter 7, have already proved amenable to manipulation, others may not be so easy. A great number of potentially useful traits have been discussed in the scientific literature (for example see Barton and Brill 1983) and apart from those already discussed here others include:

  (a) manipulating seed proteins,
  (b) improving photosynthesis,
  (c) introducing into plants the ability to fix nitrogen,
  (d) engineering tolerance to stress.

Examples (a) to (c) are most often cited, possibly because the genes encoding some of the proteins important in the individual processes have been isolated either from plants or bacteria and their individual involvement in a particular biochemical process is becoming increasingly clear. Nevertheless, engineering each of these potential traits is not without its difficulties not least in the possible deleterious effects on plant growth and productivity that changes in the biochemistry of the plant cells might result in. In examples (a) and (b), the numbers of genes involved in the processes either as individuals or as members of multigene families are likely to complicate matters and in the case of manipulating the amino acid balance of seed proteins changes in the secondary structure might change the stability and accumulation of the protein in the developing seed. Engineering plants to fix atmospheric nitrogen by reduction to ammonia has often been cited as an example of a potential benefit of plant genetic engineering. However, it is now generally appreciated that the genetic, biochemical and physiological complexity of nitrogen fixation is such that it is unlikely that this trait can be engineered into novel hosts. Nevertheless, progress may be made in increasing the efficiency of nitrogen fixation by soil bacteria, optimizing their interaction with plants and extending their host range to more agronomically important plants. Engineering stress tolerance is of great potential benefit. As discussed in Chapter 6, it is now known that plants synthesize a variety of polypeptides in response to stress and it is generally considered that these polypeptides serve to protect the plant cell from adverse environmental conditions. Nevertheless, at present our understanding of what role the proteins induced under conditions of stress might have in protecting the plant is fragmentary and it is not clear how, if they are engineered, they can protect the plant and what effect their introduction into, or over-expression in, novel hosts might be.

In order to genetically engineer a plant to improve its agronomic performance we need a clear understanding of the biochemistry and the physiology of the process that is to be manipulated as well as an ability to isolate and identify the genes whose products have a role in the individual pathway. At the level of gene cloning this may require a complete reconsideration of the techniques that are currently available for gene isolation. Indeed, completely revised technical approaches may be required to localize the individual genes important in a biochemical pathway. Certainly, the key to engineering complex traits lies in the better understanding of their genetic basis. This then can be used to identify regions of the plant genome that are responsible. The next step is to find the means to isolate those regions of the genome and pin-point the relevant genes. We will then be in a position to isolate and reintroduce them into novel hosts in order to assess whether they can confer an improved phenotype. The success of this process will demand the interaction of plant physiologists, breeders, biochemists and molecular biologists.

With this in mind we can see that gene transfer technology still has a long way to go before it can be applied to improve a variety of complex traits. Indeed in many cases we still need to be able to identify the genes that are responsible for many agriculturally useful traits. Nevertheless, in the relatively short time since gene transfer into plants has been feasible, the advances have been impressive and already, in the instance of resistance to herbicides, pathogens and pests, it has proved itself to be a feasible tool in improving current plant varieties. Coupled with advances in plant tissue culture, plant transformation is likely to become a powerful technique in plant breeding, providing novel material for breeding programmes and ways of genetically tagging genomes. There are many barriers to overcome, not least in improving our understanding of plant biochemistry, however, we have seen how gene transfer not only allows us to engineer novel traits into plants but also has the potential to help us understand the biochemical processes underlying agronomically useful traits as well as isolate the genes that are responsible for them. The potential benefit of gene transfer by non-conventional techniques is possibly as vast as the variation seen in plant form and function and the ground rules for these techniques have now been laid. In science, as in politics, it is always very dangerous to say 'never'.

References

Andre, D., Colau, D., Schell, J., Van Montagu, M. and Hernalsteens, J-P. (1986). 'Gene tagging in plants by a T-DNA insertion that generates APH (3′)II plant gene fusions', *Mol. Gen. Genet.* **204**, pp. 512–518.

Baker, B., Coupland, G., Federoff, N., Starlinger, P. and Schell, J. (1987). 'Phenotypic assay for excision of the maize controlling element *Ac* in tobacco', *EMBO J.* **6**, pp. 1547–1554.

Baker, B., Schell, J., Lorz, H. and Federoff, N. (1986). 'Transposition of the maize controlling element "activator" in tobacco', *Proc. Nat. Acad. Sci. USA* **83**, pp. 4844–4848.

Barton, K.A. and Brill, W.J. (1983). 'Prospects in plant genetic engineering', *Science* **219**, pp. 671–676.

DeBlock, M., Schell, J. and Van Montagu, M. (1985). 'Chloroplast transformation by *Agrobacterium tumefaciens*', *EMBO J.* **4**, pp. 1367–1372.

De la Pena, A., Lorz, H. and Schell, J. (1987). 'Transgenic rye plants by injecting DNA into young floral infloral tiller', *Nature* **325**, pp. 274–276.

Herman, L.M.F., Van Montagu, M. and Depicker, A. (1986). 'Isolation of tobacco DNA segments with plant promoter activity', *Mol. Cell. Biol.* **6**, pp. 4486–4492.

Klee, H.J., Hayford, M.B. and Rogers, S. (1987). 'Gene rescue in plants: A model system for "shotgun" cloning by retransformation', *Mol. Gen. Genet.* **210**, pp. 282–287.

Koncz, C., Kreuzaler, F., Kalman, Zs. and Schell, J. (1984). 'A simple method to transfer, integrate and study expression of foreign genes such as chicken ovalbumin and α-actin in plant tumors', *EMBO J.* **3**, pp. 1929–1937.

Leaver, C.J., Hack, E., Dawson, A.J., Isaac, P.G. and Jones, U.P. (1983). 'Mitochondrial genes and their expression in higher plants', in *Nucleo-mitochondrial Interactions*. Eds. Schweyen, R.J., Wolf, K. and Kaudewitz, K., pp. 269–283. Berlin, Walter de Gruyter.

Maliga, P. (1984). 'Isolation and characterisation of mutants in plant cell culture', *Ann. Rev. Plant Physiol.* **35**, pp. 519–542.

Ooms, G. and Lenton, J.R. (1985). 'T-DNA genes to study plant development: precocious tuberisation and enhanced cytokinins in *A. tumefaciens* transformed potato', *Plant Mol. Biol.* **5**, pp. 205–212.

Prosen, D.E. and Simpson, R.B. (1987). 'Transfer of ten-member genomic library to plants using *Agrobacterium tumefaciens*', *Bio/Technology* **5**, pp. 966–971.

Rothstein, S.J., Dimaio, J., Strand, M. and Rice, D. (1987). 'Stable and heritable inhibition of the expression of nopaline synthase in tobacco expressing antisense RNA', *Proc. Nat. Acad. Sci. USA* **84**, pp. 8439–8443.

Simeons, C., Alliote, Th., Mendel, R., Muller, A., Sciemann, J., Van Lijsebettens, M., Schell, J., Van Montagu, M. and Inze, D. (1986). 'A binary vector for transferring genomic libraries to plants', *Nucleic Acids Res.* **14**, pp. 8073–8090.

Teeri, T.H., Herrera-Estrella, L., Depicker, A., Van Montagu, M. and Palva, E.T. (1986). 'Identification of plant promoters *in situ* by T-DNA-mediated transcriptional fusions to the *nptII* gene', *EMBO J.* **5**, pp. 1755–1760.

Van Sluys, M.A., Tempe, J. and Federoff, N. (1987). 'Studies on the introduction and mobility of the maize *Activator* element in *Arabidopsis thaliana* and *Daucus carota*', *EMBO J.* **6**, pp. 3881–3889.

Glossary

Acetosyringone: One of the phenolic compounds released by wounded plant cells which can stimulate the transcription of the *vir* region of the Ti plasmid in the initial stages of transfer of DNA from *Agrobacterium* to the plant cell.

Agroinfection: A method for initiating viral infection by inoculating host plants with *Agrobacterium* containing viral-specific DNA within the T-DNA.

Antisense RNA: Transcribed RNA representing the opposite strand of the DNA to that of an authentic mRNA so that it can hybridize to the 'sense' RNA interfering with its expression.

Auxin: A plant growth substance which can induce the growth of roots when applied to callus.

Binary vector: A plant transformation vector based on a replicon active in both *E. coli* and *Agrobacterium* which contains the T-DNA border sequences flanking sites for cloning foreign DNA. Transfer from *Agrobacterium* to the plant cell is mediated by the *vir* region of the Ti plasmid resident in the *Agrobacterium* working in *trans*, hence the alternative name *trans* vector.

Callus: Undifferentiated mass of dividing plant cells.

CAT (chloramphenicol acetyl transferase): A procaryotic gene encoding resistance to chloramphenicol, often used as a screenable genetic marker in plant transformation.

CHMTII (Chinese Hamster metallothionein): Gene conferring tolerance to heavy metals.

cis: An activity which acts on a DNA sequence that is physically linked to its source.

Co-cultivation: A technique of transforming plant cells where protoplasts are incubated in the presence of *Agrobacterium* containing a binary or co-integrative vector engineered to contain foreign DNA.

Co-integrative vector: A plant transformation vector based on the Ti plasmid of *Agrobacterium*. Routinely, the oncogenic sequences of such vectors are removed and the remaining DNA flanked by the T-DNA borders are engineered to contain sequences which allow homologous recombination with intermediate vectors containing foreign DNA. The *vir* region of the Ti plasmid is unmodified and because it is located on the same plasmid as the foreign DNA such vectors are also known as *cis* vectors.

Concatomers: DNA fragments ligated end-to-end to form a multimer.

Crown gall: A neoplastic disease initiated by the soil bacterium *Agrobacterium tumefaciens*.

Cytokinin: A plant growth substance which when applied to callus induces the growth of shoots.

DHFR (dihydrofolate reductase): An enzyme responsible for conferring resistance to methotrexate.

Electroporation: A method by which nucleic acid is introduced into plant cells by subjecting isolated protoplasts to an electric pulse.

Enhancer: A sequence of DNA which confers enhanced levels of transcription on a gene located nearby, generally non-polar in its action.

Epigenetic: Factors effecting a plant (or cell) whose origin is not from within the plant (or cell) and has no genetic basis.

Exon: A region of the gene that encodes protein.

Explant inoculation: A technique for transforming plant cells by which tissue explants are inoculated with *Agrobacterium* containing a transformation vector engineered to contain foreign DNA.

Gene tagging: A method by which a gene can be localized by the insertion of a transposable element or the T-DNA within it or nearby.

Gene rescue: A method by which a gene can be isolated by transformation as a result of a phenotype which it can confer on a host cell.

GUS (glucuronidase): A genetic marker derived from *E. coli* which is a useful screenable marker in plant cells.

Hairy root: Neoplastic disease initiated by the soil bacterium *Agrobacterium rhizogenes*.

IAA (indoleacetic acid): An auxin.

iaaH: A gene located on the T-DNA encoding indoleacetamidehydrolase, an enzyme involved in the biosynthetic pathway of IAA.

iaaM: A gene located on the T-DNA encoding tryptophan monooxygenase, an enzyme involved in the biosynthetic pathway of IAA.

Intermediate vector: A bacterial cloning vector into which foreign DNA is engineered prior to introduction into *Agrobacterium* where it can be inserted into a co-integrative vector by recombination.

Intron: A sequence of DNA that interrupts the protein coding sequence of a gene.

ipt: A gene located on the T-DNA encoding isopentylenyltransferase considered to be important in the synthesis of cytokinins.

lux: Gene(s) encoding luciferase from either bacterial or insect source.

Mendelian inheritance: Mode of inheritance of a genetic trait expected if the gene is located in the nuclear DNA of the plant cell.

Minicells: Cells derived from the asymmetric division of mutant bacteria which although contain no chromosomal DNA do contain plasmids and the apparatus to carry out transcription and translation for a limited amount of time.

Naked DNA uptake: A method of transformation of isolated plant cells with cloned bacterial DNA which does not require *Agrobacterium* and hence circumvents the need to engineer the foreign DNA into a specialized Ti plasmid-based transformation vector.

19S RNA: The mRNA of ORFVI, the inclusion body protein gene of Cauliflower Mosaic Virus.

Northern analysis: A method by which the amount and size of a particular RNA can be assessed. RNA is isolated from tissue, denatured, fractionated by agarose gel electrophoresis under denaturing conditions and transferred to nitrocellulose paper (or membrane). The membrane is incubated in the presence of a radiolabelled DNA and any RNA corresponding to the DNA will hybridize to it and be detected by autoradiography.

nos (nopaline synthetase): A gene located on the T-DNA of some Ti plasmids which directs the synthesis of nopaline from arginine. It is a useful genetic marker as it appears to be expressed constitutively at high levels in plants.

NPTII (neomycin phosphotransferase II): A bacterial gene encoding resistance to kanamycin, a useful dominant selectable marker in plant cells.

ocs (octopine synthetase): A gene located on the T-DNA of some Ti plasmids which directs the synthesis of octopine. A useful genetic marker as it appears to be constitutively expressed at high levels in plants.

onc: Oncogenic genes, those that direct tumorous growth.

ons: A gene located on the T-DNA which is involved in the secretion of nopaline from the transformed plant cell.

Open reading frame (ORF): A region of sequenced DNA, preceded by ATG which is potentially able to encode a polypeptide.

Opine: Novel amino acid derivatives which are synthesized in plant cells transformed by *Agrobacterium*. The opine biosynthetic genes are encoded by the T-DNA.

Over-drive: Sequence located to the right of the T-DNA thought to act in *cis* to increase the levels of T-DNA transfer intermediates during plant transformation.

PEG/PLO (polyethylene glycol and poly-L-ornithine): Long-chain cations which can be used to induce the fusion of protoplasts and the uptake of nucleic acids.

Poly(A) RNA: RNA isolated from eucaryotic cells which contains a poly(A) sequence at its 3′ end, generally considered to represent mRNA.

Rolling circle replication: A method of DNA replication which proceeds by a circular intermediate producing multimeric genomes which are subsequently cleaved to produce monomeric units.

Screenable marker: A genetic marker for use in transformation which can be used to screen for changing levels of gene activity.

Selectable marker: A genetic marker for use in transformation whose product confers a phenotype which can be selected for, e.g. antibiotic resistance.

Southern analysis: A method by which the presence of a sequence of DNA can be determined within a large mixture of DNA sequences. DNA is isolated from the tissue in question, cut with restriction enzymes, fractionated by agarose gel electrophoresis, denatured and transferred to nitrocellulose paper (or membrane). The immobilized DNA is incubated with a radiolabelled DNA probe representing a specific sequence of DNA and homologous DNA, which hybridizes to the probe, and can be detected by autoradiography.

T-DNA: The region of the Ti plasmid of *Agrobacterium* which is transferred to the nuclear DNA of the plant cell and confers the tumour phenotype.

T-DNA borders: The 25 base pair, almost perfect, direct repeats of DNA sequence which flank the T-DNA and delimit the sequences which are transferred to the plant cell.

35S RNA: Greater than full length transcript of the Cauliflower Mosaic Virus genome which is considered to act both as a replicative intermediate and a polycistronic mRNA. Expressed constitutively in plant cells at high levels.

TL/TR: The two regions of T-DNA present in either octopine Ti plasmids or the Ri plasmids of *Agrobacterium rhizogenes* which can be independently transferred to the plant cell during transformation.

Totipotency: The ability of isolated plant cells to regenerate into a whole plant.

tml: A gene present in the T-DNA which regulates the size of tumours that are produced following transformation.

Transcriptional fusion: A fusion of a promoter sequence with the protein coding sequence of a gene so that the latter is not modified.

Translational fusion: A fusion of both the promoter sequence and the protein coding sequence of a gene with the protein coding sequence of another gene so that a chimeric protein is produced following translation.

Transformation: The insertion of foreign DNA into a new host.

Transient expression: The expression of foreign DNA in a cell shortly after its introduction, mediated by PEG, PLO or electroporation without selecting for stable transformation.

tzs: A gene located on the Ti plasmid encoding a protein involved in the synthesis of *trans*-zeatin, a cytokinin.

vir: The region of the Ti plasmid which encodes proteins important in the passage of T-DNA from *Agrobacterium* to the plant cell.

Index

DATE DUE

NO 26 '90			
JAN 3 2000			
JAN 07 2000			

DEMCO 38-297